T0324172

Environmental Footprints and Eco-design of Products and Processes

Series editor

Subramanian Senthilkannan Muthu, SGS Hong Kong Limited,
Hong Kong, Hong Kong SAR

More information about this series at http://www.springer.com/series/13340

Pammi Sinha
Subramanian Senthilkannan Muthu
Geetha Dissanayake

Remanufactured Fashion

 Springer

Pammi Sinha
University of Leeds
Leeds
UK

Geetha Dissanayake
University of Moratuwa
Moratuwa
Sri Lanka

Subramanian Senthilkannan Muthu
Environmental Services Manager-Asia
SGS Hong Kong Limited
Hong Kong
Hong Kong SAR

ISSN 2345-7651   ISSN 2345-766X   (electronic)
Environmental Footprints and Eco-design of Products and Processes
ISBN 978-981-10-0295-3   ISBN 978-981-10-0297-7   (eBook)
DOI 10.1007/978-981-10-0297-7

Library of Congress Control Number: 2015958548

Printed on acid-free paper

This Springer imprint is published by SpringerNature
The registered company is Springer Science+Business Media Singapore Pte Ltd.

Preface

This book is developed from our research backgrounds and experiences. Sinha's background in fashion design process and her research experiences within the Tanzanian fashion and textiles design community initiated in 2007 led to her sustained interest in issues regarding managing the rising volumes of second-hand and discarded fashion and textiles with particular interest on West Africa; subsequent funding facilitated a Ph.D. studentship to examine sustainable fashion design strategies conducted by Dissanayake. Dissanayake's background and expertise in factory production and garment manufacture were fully utilized to examine the processes at several upcycled fashion design studios, waste textile collection companies and a mass-market retailer, forming the basis of her Ph.D. Through discussions with Muthu, an expert in textiles and clothing sustainability and life cycle assessment, we realized that there was much more to explore about the remanufacture concept for the fashion industry, and so we three authors set out to write this handbook; any errors, oversights or omissions are our own.

This book considers the impacts on and implications for the fashion system, if the mass market were to adopt remanufactured fashion as a design and manufacture strategy. Managing textile waste is a major sustainability issue for any country. Following the waste hierarchy, conversion of waste to a useful product is regarded as one of the more sensible options to tackle this issue as energy and resources used in the product's manufacture are retained and useful life is prolonged. Remanufacturing retrieves a product's inherent value when the product no longer fulfils the user's desired needs. Using discarded clothes in a remanufacturing process contributes to resource conservation by reducing the percentage of clothing discarded as waste, which is still either dumped at landfill, or incinerated, posing many environmental issues.

Remanufactured fashion, a sustainable waste management strategy, deals with discarded garments by amending and converting them into useful products for retail at mass-market prices. The focus on the mass market is to engage large volumes of consumers in this approach. There are arguments to be made about the proliferation of many small studios achieving the same volume (through several niche markets), but our reservations are (i) remanufactured fashion would remain an "alternative" approach and not considered as part of mass-marketing retailer's closed loop approach to fashion design and (ii) the (often detrimental) impacts of market competition on small design studios. There has been scant research investigating what is actually involved in the fashion remanufacturing process and how the process could be upscaled to the mass market in order to achieve greater environmental gains than that which is currently achieved. This book aims to fill this literature gap and will examine all aspects pertaining to the concept and applications of remanufactured fashion through six chapters:

Chapter 1—the term "remanufacture" within the literature is examined, and a definition for the remanufactured fashion arena is presented.

Chapter 2—the concepts of closed loop production, reverse supply chains and logistics are examined and placed into the contexts of the fashion industry.

Chapter 3—current business models and strategies adopted by sustainable fashion are examined and compared with the remanufactured fashion process in order to identify the implications for design, manufacture and retail.

Chapter 4—case studies are presented for fashion design and retailing companies to examine the extent to which the remanufacture process is practiced.

Chapter 5—a conceptual system for remanufactured fashion is presented and examined for drivers and obstacles with the implications on new technologies, marketing opportunities and business strategy.

Chapter 6—sustainability implications are examined for retailing and certification.

The authors are indebted to the Sustainable Consumption Institute, University of Manchester, Tesco and the Dorothy Hodgkin Postgraduate Award for the funding that enabled the Ph.D. study and also to the fashion designers and the textile waste collection companies who gave such generous access to their studios and time to talk about their processes.

Contents

1 The Remanufacturing Industry and Fashion 1
  1.1 Introduction 1
  1.2 The Remanufacturing Industry 2
  1.3 The Concept of "Remanufacture" 5
  1.4 A Definition of Remanufactured Fashion 5
  1.5 Upcycle or Remanufacture 6
  1.6 Labelling of Remanufactured Clothing 7
  1.7 Conclusion 7
  References 8

2 Closed Loop Systems and Reverse Logistics 11
  2.1 Introduction 11
  2.2 Closed Loop Systems, Reverse Logistics—Definitions
     and Differences 12
    2.2.1 Closed Loop Versus Reverse Logistics 13
  2.3 The Process of Reverse Logistics 13
  2.4 Requirements and Challenges of Closed Loop Systems
     of Fashion Products 13
  References 15

**3 The Remanufactured Fashion Design Approach
  and Business Model** 17
  3.1 The Business Model Concept 17
  3.2 Significance of the Business Model Concept 17
  3.3 Business Models and Sustainability 18
  3.4 Business Models Within a Globalized Fashion Industry 18
  3.5 Sustainable Fashion Business Models 19
    3.5.1 Incentive Green Business Model 19
    3.5.2 Life-cycle Green Business Models 19
  3.6 An Examination of the Remanufactured Fashion
     Business Model 20

3.7   The Remanufactured Fashion Business Model 22
       3.7.1   Value Creation 22
       3.7.2   Value Architecture 23
       3.7.3   Revenue Model....................................... 24
3.8   Implications for the Design and Manufacture
       of Remanufactured Fashion 24
3.9   Conclusion 28
References 28

4   Examples and Case Studies 33
    4.1   Introduction 33
    4.2   Overview of the Fashion Remanufacturing Process. 33
           4.2.1   Company A.. 33
           4.2.2   Company B.. 35
           4.2.3   Company C.. 36
           4.2.4   Company D.. 38
           4.2.5   Company E.. 40
    4.3   Implications for Mass Manufacturing 41
           4.3.1   Process Input 42
           4.3.2   Cutting Operation................................. 42
           4.3.3   Garment Assembly................................. 42
           4.3.4   Quality Standards................................. 43
           4.3.5   Distribution and Retailing 44
    4.4   Conclusion 44

5   Systems Requirements for Remanufactured Fashion
    as an Industry 45
    5.1   Introduction: A Comparison Between Conventional
           and Remanufactured Fashion Design Processes 45
    5.2   Remanufactured Fashion as a Reverse Supply System 47
    5.3   The Current Fashion System 53
    5.4   A Conceptual System for Remanufactured Fashion 58
    5.5   Marketing and Strategic Considerations
           for the Remanufacturing Company 61
    5.6   Conclusion 67
    References 68

6   Issues Raised for Sustainability Through Remanufactured
    Fashion 73
    6.1   Retailing of Remanufactured Fashion 73
    6.2   Eco-Labels 74
    6.3   Remanufactured Certificates 76
    6.4   Conclusions and Perspectives.............................. 80
    References 81

Chapter 1
The Remanufacturing Industry and Fashion

1.1 Introduction

Remanufacturing is the process of disassembling, cleaning, inspecting, repairing, replacing and reassembling the components of a part or product and differs from repair or reuse as quality and lifespan of the product should be as a new product (APSRG 2014a). As a process, remanufacture is more complicated than traditional manufacture because of the variability in quality and quantity of returned goods.

Literature examining remanufacturing practices within the automotive and electronics industries are growing, however, there is little examination of the concept and practise of remanufacture within the fashion industry. Reservations have been raised about the potential for remanufacturing to be adopted by the fashion industry because of rapid changes in aesthetics and potential degradation of fabrics. We contend that the fashion industry is a unique manufacturing industry, full of contradictions; it is simultaneously highly symbolic (satisfying the consumer's desire to express their personal values and beliefs) and uniform (satisfying the consumer's need to "fit into" a group and acknowledge modernity of thought). As such, the term "remanufacture" for fashion industry needs to take this into account.

Remanufacture is regarded as being high in the waste hierarchy. The APSRG (2014b) outlined their waste hierarchy which (from highest to lowest) is as follows: (i) prevention, (ii) minimization, (iii) remanufacture, (iv) reuse, (v) recycle, (vi) energy recovery and (vii) disposal. Where (the current) linear manufacturing practices are material, resource and energy intensive, with ensuing environmental issues, remanufacture is part of the circular economy (Ellen MacAruther Foundation 2013), using less resources, materials, energy and water and producing fewer greenhouse gases.

© Springer Science+Business Media Singapore 2016
P. Sinha et al., *Remanufactured Fashion*, Environmental Footprints
and Eco-design of Products and Processes, DOI 10.1007/978-981-10-0297-7_1

1.2 The Remanufacturing Industry

Current remanufacture practices (business models) have been identified that have been labelled as follows: (i) fee for service, (ii) pure leasing, (iii) incentivized return and reuse and (iv) third-party remanufacturing (APSRG 2014b). The models centre on the mechanisms to manage the product at the end of life, and if a product is remanufactured at the end of its life or end of contract. These models are outlined and compared in Table 1.1. It has been also found that OEM's either embrace or shun third-party remanufacturers largely due to the IP issues involved. Remanufactured products may be an infringement of patent laws raising problems in outsourcing remanufacture capacity. Some companies have overcome this by purchasing the original design from the OEM to overcome the reverse engineering that would otherwise be involved and patent laws.

The potential social, economic and environmental benefits of remanufacture have led some authors to describe this as a "triple win" (APSRG 2014b). The US remanufacturing industry is the largest and most established globally. From the last collected figures in 2011, it has been valued at $43 billion (£27.4 million) and amounts to 2 % sales of all manufactured goods (USITC 2012). The most active sectors for remanufacture are automotive and aerospace. 25 % of the remanufactured goods were from SME's, and this constituted 17 % of their exports. The American remanufacturing industry supports over 180,000 full-time jobs of which 36 % were in the SME's. Moreover, as 34–45 % of the production costs are on labour, remanufacturing presents an opportunity for the growth of various levels of skilled jobs. The American remanufacturing industry identified that the workforce could be categorized as follows: 50 % skilled, 30 % semi-skilled, 10 % professional and 10 % unskilled.

The EU is the next largest area of remanufacturing, and estimates are that it is worth €40 billion (£31.8 million) supporting 300,000 jobs. The areas of highest remanufacturing activity are considered to be in UK and Germany; the UK remanufacturing industry is valued at £2.4 billion. Although the study could find no empirically based recorded data in China, they did identify the concerted government's effort on developing the remanufacturing industry in China (particularly in automotive) through the use of national policies, funding and regulations, e.g. National Development and Reform Commission which supports and funds remanufacture through the 12th Five-Year Plan 2011–2015. This was similar in other BRIC countries such as Brazil and India, where the efforts to develop the remanufacturing industry were also unfortunately hampered by confusing the term remanufacture with repair or recondition. The lack of a universally accepted legal definition of the term "remanufacture" and the remanufacture process has been cited as a principal barrier to international trade (USITC 2012; APSRG 2014b). A universally accepted definition of the term "remanufacture" therefore is an important element to help policy-makers and governments develop routes to a circular economy.

Table 1.1 Comparison of remanufacturing business models

	What is paid for	Who owns what	Advantages to the company	Advantages to customer	Issues raised/to note
Fee for service (FFS)	Customer pays a term-based fee for a performance output rather than the product	The OEM retains ownership of the product and the system—it installs, maintains, services, repairs, replaces at the end of life and reclaims at the end of contract	OEM may operate a remote management system for that remotely monitors usage data—they can advise the customer on ways to maximize usage or have prior notice of maintenance issue thus advantages include	Guarantee of product functionality, regardless of warranty—all inclusive pricing means simplicity of service, eliminating the need to research for appropriate people to provide service for the product. (Examples are the photocopiers in most businesses)	Originally called "product service system"
			(1) Reduce waiting time for maintenance to a minimum (causing little/no delay for the customers)		
			(2) Establishes a long-lasting, deeper relationship with the customer to serve them better—thus happy customer		Is the most well-developed system and has the most chance of becoming a robust system for remanufacture, specially of high-value products that have a high degree of IP
			(3) Finally, as the OEM owns the product, the reclamation of the product gives them (i) 100 % access to materials for remanufacturing process, (ii) guarantee of revenue and (iii) predictability of revenue		Reclamation of the product at the end of the contract could displace the waste collection agencies

(continued)

Table 1.1 (continued)

	What is paid for	Who owns what	Advantages to the company	Advantages to customer	Issues raised/to note
Pure leasing (PL)	Customer pays for a short-term contract to hire the product—not the service. Service comes at an extra cost. Customer owns the product until the end of the contract period after	Company part owns the product until the end of contract—if customer wants to buy, then the product is no longer the company's. All services are owned by the company	(i) Some advantages in potential for long-term relationships.	"Try before buy" approach. The customer does not have to buy the product from the company hired from—can try out then go to another retailer	Similar to hiring/leasing
			(ii) Looser contract means that security of revenue not as high and predictability lowered		
			(iii) Less access and control of product than FFS		Best suited for high capital industrial items or those that are not needed all the time
Incentivized return and reuse (IRR)	Customer pays for the product plus a surcharge fee that is returned upon return of the product at the end of life	Customer owns the product	Potential to recapture product for remanufacture but requires customer to bring back	Outright ownership of the product can modify or transfer ownership of the product or dispose of it	Opportunity for relationships with waste collectors to be developed and also with government links
			Follows the traditional linear model most closely	One upfront cost (not repetitive)	Potential for remanufactured products to be exported out of the UK (e.g. buses and taxis currently exported)
Third-party remanufacturing (TPM)	Never owns the product but owns the remanufacture service	OEM retains control of the product	Payment for remanufacture	Outsourced production and the expertise and time involved	TPM-O-like tradition manufacturer—is not the OEM
Outsourced (TPM-O)					Some businesses do not understand the term "remanufacture" or do not want to use are remanufacturer
Systems approach (TPM-S)	Remanufacture is one element of a system owned by the TPM-S company	OEM retains control of product; TPM-S retains control of the system			

Source APSRG (2014b)

1.3 The Concept of "Remanufacture"

The term "remanufacture" was published in a scientific context by Lund (1984) who described it as "... an industrial process in which worn-out products are restored to like-new condition. Through a series of industrial processes in a factory environment, a discarded product is completely disassembled. Useable parts are cleaned, refurbished, and put into inventory. Then the product is reassembled from the old parts (and, where necessary, new parts) to produce a unit fully equivalent and sometimes superior in performance and expected lifetime to the original new product" (Lund 1984, p. 19). This approach to sustainable manufacturing has been practised and examined within the engineering and electrical and electronics industries—most often within automotive and aerospace, where this is a common practice. Indeed, the automotive industry in the USA has an association dedicated to this practice: the Automotive Parts Remanufacture Association (APRA) where they use the term "rebuild" interchangeably with the term "remanufacture" (www.apra.org). As a result of research and workshops with industrialists, Ijomah et al. (2007) proposed the following definition of "remanufacture" as "The process of returning a used product to at least OEM original performance specification from the customers' perspective and giving the resultant product a warranty that is at least equal to that of a newly manufactured equivalent." A product could be remanufactured with or without its original product identity (Gray and Charter 2008). If the new product is an assembly of parts from the original product, the original identity of the product would remain with the new product. If the parts are assembled from different products, the new product would lose its original product identity. Remanufacturing process steps (e.g. collection, inspection, cleaning, repairing/reworking and manufacturing) could be put in a different order, or some steps could even be omitted, depending on the product type, product design, remanufacturing volume, etc. (Sundin 2004).

1.4 A Definition of Remanufactured Fashion

Building on this definition, the term "remanufactured fashion" may be defined as "the process of remaking used clothes into *new* clothing that is at least equal to if not better than the original manufacture specifications from the consumer's perspective and the brand and garment labels attached to the garment will indicate that the garment has been remanufactured to at least equal quality to that of a newly manufactured equivalent." The remanufactured product may or may not continue the original identity, design or the functionality of the product. For example, a dress could be remanufactured into a skirt, or a trouser could be remanufactured into a coat. The nature of fashion industry is such that a particular trend would not last for long. Designs keep changing as the trend changes, and a particular design from a previous season might not be continued for the upcoming season. In this context, remanufacturing the original product without losing the original design or product identity would not be a feasible strategy. Therefore, the concept of remanufacturing for the fashion industry is somewhat different from other remanufacturing industries in terms of approach, process and end product.

1.5 Upcycle or Remanufacture

There is much confusion about the term "remanufacture" as a recent report by the UK government attests "there is no universally accepted definition of remanufacturing" (APSRG 2014a, b, p. 1) and cites this lack of definition as a problem in developing legal frameworks for international trading and consumers' confidence. The report cites a number of alternate terms that are confused with "remanufacture" such as "repair" and "refurbishment" (APSRG 2014b, p. 2). In the fashion industry, it is the term "upcycle" that is most commonly used interchangeably, and the distinctions between them have been poorly explained. The similarity between the two terms is that both are strategies to avoid wasting materials by using them to design products of at least equal to if not higher value than the original product held. From the literature reviewed, the differences appear to be the design goal or strategy, the process approach, product end use or function, the material input and the need for a warranty, as illustrated in Table 1.2.

Table 1.2   Differences between the terms "upcycle" and "remanufacture" (*source* Authors)

	Upcycle	Remanufacture
Goal/design strategy	Achieve a higher value at retail than the original product would (Pilz in Thornton 1994; McDonough and Braungart 2002)	Achieve an "as good as new" product that is at least equal to if not better than the original OEM specifications that can be sold at as near or slightly under the original price (Lund 1984; Ijomah et al. 2007)
Process approach	Craft, individual and possibly unique product requiring (often) manual intervention (Vermeer 2014; Upcycle magazine 2009)	Industrial process carried out in factory environment (He 2015; Goodall et al. 2014; Hazen et al. 2012; Steinhilper and Hieber 2001; Lund 1984)
Product end use or function	Can serve a completely different function or end use from original use (Sung 2015; Cassidy and Han 2013; Upcycle magazine 2009; Pilz 1994)	Should serve the same function or end use as the original (Hatcher et al. 2014; Lund 1984)
Material input	May or may not have been used (i.e. the materials may be spare for the production line) and, therefore, may or may not be faulty (Sung 2015)	Have been used, they may be worn out in parts, or destined for waste if not used (Lund 1984; Hatcher et al. 2014)
Warranty	Not required	Quality indicator is necessary (such as a warranty) both to attest to the "good as new or better" quality and to differentiate from a "new" product (Ijomah 2007; Automotive Parts Remanufacture Association; Lund 1984)

1.6 Labelling of Remanufactured Clothing

The lack of universal clarity regarding definitions for the term "remanufactured fashion" compound the problems regarding developing a warranty for remanufactured fashion. The use of a warranty within the fashion industry is a complex legal issue that is more typically seen with regard to retailing and licensing of brands, sales and returns policies but sometimes also as part of a product returns policy (e.g. outdoor activity clothing retailers or manufacturers who may resurface or recycle appropriately labelled garments, e.g. Patagonia, Páramo). While brand names and garment labels suggest a quality indicator, they do not necessarily indicate that the garment is remanufactured (although the marketing and promotions surrounding them may do so). Given the environmental significance of developing remanufactured fashion, this indicates a potential to create a label for the remanufactured fashion process to both claim environemental benefits and also verify and authenticate that the garment on the shop floor is made from clothing that may have been previously been worn but is of at least equal if not higher manufacturing specifications to the original. As yet we have not discovered an eco-label for remanufactured fashion but Chap. 3 outlines the current eco-labels that are within recycled textiles and considers the potential for a remanufactured fashion eco-label and Chaps. 5 and 6 further examine the issues raised in developing some form of warranty for remanufactured fashion.

1.7 Conclusion

In the 1970s, the concept of product return management became an issue related to sustainable development, and the recovery practices were mandated through environmental legislation (Mollenkopf et al. 2007). Waste management strategies in the fashion industry have developed in response to the EU Waste Framework Directive 2008/98/EC which encourages the application of the "waste hierarchy"—preference to eliminate waste at source, then, to reduce, reuse or recycle waste, and if impossible or impracticable, disposal in a responsible manner. The central notion is to decouple waste growth from economic growth. It is recognized that a mix of options may be needed to arrive at the most balanced environmental, social and economic solution. As the fashion industry experienced a rapid growth in production and consumption during last few decades and consumers have adopted to a fast fashion culture, the waste generated has increased as a consequence. In this context, direct reuse and recycling strategies were not proved to be sufficient to treat all wastes produced by the fashion industry, and therefore, remanufacturing has been recognized as a sustainable alternative to divert massive wastes from landfills. Approach to remanufacturing in the fashion industry is substantiated as fashion consumption is based on trend rather than requirement, and clothing material quality often outlasts the period of time that a product is 'on-trend'. Recovering the intrinsic value of material and using them in a reman-ufacturing process could be described as a cradle-to-cradle approach (Braungart and McDonough 2009), where materials are recovered and used again instead of being down-cycled into low-value products. Remanufacturing is recognized as the most desirable end-of-life product management option for several industries as it represents a higher form of reuse that focuses on value-added recovery (Guide 2000), results in lower energy and virgin material use, and creates new employ-ments through new business models (Nasr and Thurston 2006; Michaud and Llerena 2006; Pagell et al. 2007). Remanufacturing in the fashion industry is still at a nascent stage; however, the sustainable vision of the fashion industry needs new approaches to remanufacturing that go beyond traditional way of designing and making fashion clothing.

References

APSRG (2014a) Remanufacturing: towards a resource efficient economy. The All-Party Parliamentary Sustainable Resource Group, Mar, http://www.policyconnect.org.uk/apsrg/research/report-remanufacturing-towards-resource-efficient-economy-0

APSRG (2014b) Triple win: the economic, social and environmental case for remanufacturing. Dec 2014, http://www.policyconnect.org.uk/sites/site_pc/files/.../apsrgapmg-triplewin.pdf

Automotive Parts Remanufacture Association (APRA) www.apra.org

Braungart M, McDonough W (2002) Cradle to cradle. Remaking the Way We Make Things, Vintage

Braungart M, McDonough W (2009) Cradle to cradle: re-making the way we make things. Vintage Books, London

Cassidy D, Han S (2013) Upcycling fashion for mass production. In: Gardetti MA, Torres AL (eds) Sustainability in fashion and textiles: values, design, production and consumption. Greenleaf Publishing, pp 148–163

Ellen MacAruther Foundation (2013) Towards the circular economy, opportunities for the consumer goods sector, http://www.ellenmacarthurfoundation.org/publications/towards-the-circular-economy-vol-2-opportunities-for-the-consumer-goods-sector

Goodall P, Rosamund E, Harding J (2014) A review of the state of the art in tools and tech-niques used to evaluate remanufacturing feasibility. J Cleaner Prod 81:1–15, http://dx.doi.org/10.1016/j.jclepro. Elsevier Ltd., 14 June 2014

Gray C, Charter M (2008) Remanufacturing and product design: designing for the 7th genera-tion. Available at: http://cfsd.org.uk/Remanufacturing%20and%20Product%20Design.pdf

Guide VDR (2000) Production planning and control for remanufacturing: industry practice and research needs. J Oper Manag 18:467–483

Hatcher GD, Ijomah WL, Windmill JFC (2014) A network model to assist 'design for remanu-facture' integration into the design process. J Clean Prod 64:244–253 (Elsevier Ltd.)

Hazen BT, Overstreet RE, Jones-Farmer LA, Field HS (2012) The role of ambiguity tolerance in consumer perception of remanufactured products. Int J Prod Econ 135:781–790 (Elsevier Ltd.)

He Y (2015) Acquisition pricing and remanufacturing decisions in a closed-loop supply chain. Int J Pro Econ 163:48–60 (Elsevier Ltd.)

Ijomah WL, McMahon CA, Hammond GP, Newman ST (2007) Development of design for remanufacturing guidelines to support sustainable manufacturing. Robot Comput Integr Manuf 23:712–719

Lund RT (1984) Remanufacturing. Technol Rev 87(2):19–23, 28–29 (MIT)

Michaud C, Llerena D (2006) An economic perspective on remanufactured products: industrial and consumption challenges for life cycle engineering. In: proceedings of 13th CIRP international conference on life cycle engineering. Leuven, May 31–June 2 2006, pp 543–548

Mollenkopf M, Russo I, Frankel F (2007) The returns management process in supply chain strategy. Int J Phys Distrib Logistics Manag 37(7):568–592, http://dx.doi.org/10.1108/09600030710776482

Nasr N, Thurston M (2006) Remanufacturing: a key enabler to sustainable product systems. Rochester Institute of Technology

Pagell M, Wu Z, Murthy NN (2007) The supply chain implications of recycling. Bus Horiz 50.133–143

Pilz in Thornton K (1994) Salvo in Germany—Reiner Pilz, SalvoNEWS, 12 Oct 1999, p 14 (http://www.salvoweb.com/files/salvonews/sn99v3.pdf). Accessed 17 July 2015

Steinhilper R, Hieber M (2001) Remanufacturing-the key solution for transforming downcycling into upcycling of electronics. In: Proceedings of the 2001 IEEE international symposium on electronics and the environment, pp 161–166

Sundin E (2004) Product and process design for successful remanufacturing. Published doctoral dissertation. Linköping's University, Sweden

Sung K (2015) A review on upcycling: current body of literature, knowledge gaps and a way forward. In: Part I, ICEES 2015: 17th international conference on environmental and earth sciences, vol 17, no 4, Venice, Italy, 13–14 Apr 2015

Upcycle magazine (2009) What is upcycling? http://www.upcyclemagazine.com/what-is-upcycling. Accessed 14 July 2015

USITC (2012) Remanufactured goods: an overview of the United States and global industries, markets, and trade, United States international trade commission. Accessed at https://www.usitc.gov/publications/332/pub4356.pdf

Vermeer D (2014) 7 upcycling companies that are transforming the fashion industry, http://daniellelvermeer.com/blog/upcycled-fashion-companies. Accessed 14 July 2015

Chapter 2
Closed Loop Systems and Reverse Logistics

2.1 Introduction

Currently, textile and clothing sector is facing a severe environmental threat in the form of the following issues from the entire life cycle of textiles and clothing products:

- Depletion of non-renewable resources;
- Massive energy, chemicals and water demands;
- Huge GHG emissions;
- Human Toxicity;
- Severe environmental pollution and damage;
- Huge waste generation;
- Fast fashion cycles;
- Enormous textile waste at the end of life;
- Limited landfill space (Muthu 2014a; Gardetti and Muthu 2015; Muthu 2014b).

Apart from the above-mentioned environmental issues, there is an umpteen number of social and economic issues bundled with the entire life cycle of clothing products. To address all these issues to a decent level, at least a product has to be used till its functional limit permits (of course, use phase emissions need to be taken into consideration) and the other solution which can help the clothing sector to address this threat is the option of closed loop system with reverse logistics (RL). This concept of closed loop systems thinking is getting familiarized in the textiles and fashion sector and certainly it is the need of the hour.

Sustainability issues of the textiles and fashion sector are one side and the other side, there is a stringent regulation on the industrial waste management intensified by the growing environmental and social impact concerns of waste management across the globe. Both of these issues are equal to two sides of the same coin and the best way to tackle both is RL coupled with closed loop supply chain/closed loop systems thinking.

© Springer Science+Business Media Singapore 2016
P. Sinha et al., *Remanufactured Fashion*, Environmental Footprints
and Eco-design of Products and Processes, DOI 10.1007/978-981-10-0297-7_2

The concept of closed loop is to avoid and conserve the waste or otherwise it will be landfilled. It considers waste and discarded materials as a resource to begin the life cycle of a new product. Simply speaking, it aids to begin the life cycle of another product from the waste.

A textile product at its end of life can go to one of the following destinations (Ferguson and Gilvan 2010):

1. Landfilling—dump into landfills; needless to explain the consequences of landfill dumping as it is well known and discussed elsewhere;
2. Incineration—Burning of the waste with and or without energy recovery option and its merits and demerits are well discussed in the literature;
3. Recycling—processing the product (or waste) and converting it to a new product for the same or for different purpose;
4. Resale of the product (as it is without any further process)—for secondary products' market;
5. Internal reuse
6. Remanufacturing or refurbishing—a process with adds value to the product and it is the option with maximum profitability among all the 6 destinations mentioned here. Remanufacturing is defined as a process of restoring used products to a "new-like" condition (Melina and Siu 2011).

This chapter will deal with the concepts of RL, closed loop systems and their implications in fashion and textile sector. To begin with, this chapter clarifies the definitions and the differences between the closed loop systems and the RL as there is always a misconception on these terminologies exists.

2.2 Closed Loop Systems, Reverse Logistics—Definitions and Differences

A closed loop system is a system where products and their associated components are designed, produced, consumed and maintained, thereby to circulate within society for as long as possible, with the maximum product longevity coupled with the maximum efficient utilization of water, energy, chemicals and other resources throughout the entire life cycle along with the minimum waste production and least environmental damage and pollution.

In closed loop supply chains, along with the flow in forward direction (from suppliers to end customers), there will also be flows of products back to manufacturers from end customers (Guide et al. 2003). Chief focus lies in closed loop supply chains on bringing the products back from customers and recovering them or reusing them (whole or the parts of the products), thereby adding value to the end product (Guide and Van Wassenhove 2009).

Reverse logistics (RL) is the converse phenomenon of traditional or forward logistics (Beamon 1999). RL is defined as a process where a manufacturer accepts previously shipped products from the point for consumption for possible recycling and remanufacturing (Carter and Ellram 1998; Dowlatshahi 2000).

2.2.1 Closed Loop Versus Reverse Logistics

Reverse logistics is the process of moving or transporting goods from their final destination for the purpose of capturing value or for the proper disposal. It involves the processes for sending new or used products "back up stream" for repair, reuse, refurbish, resale, recycling or scrap/salvage. Closed loop supply chains are designed and managed to explicitly consider both forward and reverse flow activities in a supply chain (John et al. 2008; Carecprogram 2010).

2.3 The Process of Reverse Logistics

Reverse logistics demands careful design, planning and control activities. The RL network involves the collection of used products, consolidation, inspection and sorting, and transportation of those products for various recovery options (Ferguson and Gilvan 2010; Melina and Siu 2011; Guide et al. 2003; Guide and Van Wassenhove 2009). Various important activities involved in the process of RL are mentioned below (Guide et al. 2003; Guide and Van Wassenhove 2009):

- Acquisition of used products;
- Movement of products from the point of use to the point of disposition;
- Testing, sorting and disposition to determine the product's actual condition and to decide the most economically viable reuse option;
- Refurbishing to enable the most economically viable and attractive option from one of the following: direct reuse, repair, remanufacture, recycle or disposal;
- Remarketing of refurbished goods;
- Distribution of refurbished goods;

2.4 Requirements and Challenges of Closed Loop Systems of Fashion Products

Closed loop systems with RL of textiles and clothing products would be possible if the following conditions are met:

- **Design of Products**: Products need to be designed in such a way that they can be disassembled to separate components if they need to be paired or if some malfunctioned parts need to be replaced. Design of products must enable a redesign too and also they must aid the recycling of separate components at the end of the product's life. Possibly, the individual parts need to be technically recycled (Zippers, buttons, etc. from a garment at the end of life) and if not, they must be biodegraded and composted (For instance, cotton or wool).

- **Chemical-free and safe recycling**: Products need to be produced without hazardous chemicals and dangerous substances which will enable the safe recycling at industrial scale. This will aid safe biodegradation and composting too.
- **Consumer's Willingness**: Products must be returned by their customers after the useful life time to the collection points. Followed by which, the products need to be transported to larger facilities for effective sorting and recycling. This would be possible if consumers are willing and encouraged to send their used/discarded textile products to the designated collection points, with or without an economic gain.
- **Collaborative Efforts**: This whole process of RL and closed loop system is possible only if there exists a collaborative effort between:
  - Successfully running business-to-business and cross-sector collaborations in society that enable products to be repaired, refurbished and redesigned (for the extension of life time of products under question);
  - Service agencies which can provide products for rent or as second-hand items;
  - Logistics, facilities and services to support the RL operations such as collection, sorting, and recycling of components and materials, so that these may be incorporated into new products, thus giving them new life (Green Strategy 2015).

There are currently many challenges to be faced by the textile and fashion sector to be successful in the closed loop systems. There are many uncertainties in terms of the following factors:

- Amount of waste materials recycling recovery and arrival time;
- Quality of the recycled products vis-à-vis the virgin ones;
- Demand of manufactured goods;
- Remanufacturing cost;
- Remanufacturing operations—complexities in logistics, inventory, production planning, etc. (Xia et al. 2011).

In addition to the above, there are some more specific textile industry-based challenges such as (Green Strategy 2015):

- Blending of various fibres for functionality;
- Mixed scenarios of global, regional and local loops;
- Materials, components and chemicals for one single product are sourced from the different parts of the world. In such case, where do we put the degraded matter of biodegradable products?
- Management of separated components after sorting. How one can proceed with imparting a new functional life for a metal zipper or a plastic button after being separated from a jacket for example?
- Supply chain transparency;
- Conscious consumer behaviour;
- Factory working conditions, etc.

References

Beamon B (1999) Designing the green supply chain. Logistics Inf Manag 12(4):332–342

Carecprogram.org (2010) Chapter 15 managing reverse flows in the supply chain, obtained from: http://www.carecprogram.org/uploads/events/2010/1st-CFCFA-Logistics-Training/Ch15-Supply-Chain-Managing-Reverse-Flows.pdf. Accessed 1 June 2015

Carter CR, Ellram LM (1998) Reverse logistics: a review of the literature and framework for future investigation. J Bus Logistics 19:85–102

Dowlatshahi S (2000) Developing a theory of reverse logistics. Interfaces 30:143–155

Ferguson ME, Gilvan CS (2010) Closed-loop supply chains. Taylor and Francis Group LLC, Boca Raton

Gardetti MA, Muthu SS (2015) Sustainable apparel? Is the innovation in the business model?—The case of IOU project. Text Clothing Sustain 1:2

Green Strategy (2015) "Closing the loop" in the fashion industry, obtained from: http://www.greenstrategy.se/closed-loops-fashion-textile-industry-definition-and-challenges-2/. Accessed 11 June 2015

Guide VDR, Van Wassenhove LN (2009) The evolution of closed-loop supply chain research. Oper Res 57(1):10–18

Guide VDR, Harrison TP, Van Wassenhove LN (2003) The challenge of closed-loop supply chains. Interfaces 33(6):3–6

John C, Langley C, Gibson B, Novack R, Bardi E (2008) Supply chain management: a logistics perspective. Cengage Learn

Melina C, Siu L (2011) Sustainability in fashion supply chains by closed-loop logistics systems case study: Min Boutique Group, Molde—Norway. MSc in Logistics Thesis, Molde University College, Molde, 24 May 2011

Muthu SS (2014a) Assessing the environmental impacts of textiles and the clothing supply chain. Woodhead Publishing, UK

Muthu SS (2014b) Roadmap to sustainable textiles and clothing. Environmental and social aspects of textiles and clothing supply chain (Preface, V)

Xia W-h, Jia D-y, He Y-y (2011) The remanufacturing reverse logistics management based on closed-loop supply chain management processes. Procedia Environ Sci 11:351–354

Chapter 3
The Remanufactured Fashion Design Approach and Business Model

3.1 The Business Model Concept

The term "business model" is generally accepted, but there is no singular definition. Two notable descriptions, however, form the basis of much research within the business model area:

- A business model "articulates the logic and provides data and other evidence that demonstrates how a business creates and delivers value to customers. It also outlines the architecture of revenues, costs and profits associated with the business enterprise delivering the value" (Teece 2010, p. 173) and
- "a business model describes the rationale of how an organisation creates, delivers, and captures economic, social, and other forms of values" (Osterwalder and Pigneur 2010).

3.2 Significance of the Business Model Concept

According to Wirtz et al. (2015), the scientific use of the term "business model" can be traced to Bellman et al. in (1957). Locating the place of the business model within academic literature has been difficult as noted by Teece (2010) and the concept of "business model" has been criticized because of the imprecision of the term and the resultant lack of developing well-researched company strategy and competitive advantage (Porter 2001). Indeed, a business model is "conceptual, rather than a financial, model of a business" and "outlines the business logic required to earn a profit (if one is available to be earned) and, once adopted, defines the way the enterprise goes to market" (Teece 2010, pp. 173–174). However, the static concept of the business model has developed from a heuristic, descriptive and operational perspective (often in the realms of information technology and business

© Springer Science+Business Media Singapore 2016
P. Sinha et al., *Remanufactured Fashion*, Environmental Footprints and Eco-design of Products and Processes, DOI 10.1007/978-981-10-0297-7_3

process modelling) to become a management tool (Wirtz et al. 2015) with a more strategic approach that seeks to examine elements of competitiveness and innovation. This is due to the structured and holistic approach to understanding a firm's organizational structure, capabilities, target markets and value propositions. As noted by Storemark and Hoffmann (2012), business model literature may be regarded as an "interesting alternative to the positioning (Porter 1985), resource based (Barney 1991) and relational (Dyer and Singh 1998) schools of strategy (Warnier et al. 2004)" (Storemark and Hoffman 2012, p. 36).

3.3 Business Models and Sustainability

"Green business models" have been defined as "business models which support the development of products and services (systems) with environmental benefits, reduce resource use/waste and which are economically viable. These business models have a lower environmental impact than traditional business models" and highlight their aim to create economic and environmental win–win benefits for both the supplier and the customer (FORA 2010). Two main types of green business models have been distinguished: incentive models that create incentives for customers to use resources more efficiently and life-cycle models that focus on the greening of a companies' value chain (Nordic Innovation 2012). Although growth in interest in business models and value creation within eco-innovation literature have been noted, it has been recognized that the most important element of a business model for customer and company is economic value for eco-innovation to take place (Beltramello et al. 2013).

3.4 Business Models Within a Globalized Fashion Industry

To explain the complexities within a firm and its value chain (activities, relationships and links between companies along the value chain), a perspective view is necessary. Within the globalized fashion industry, these following perspectives have been taken:

- retailing and marketing perspective such as couture, luxury, designer, premium and fast retailers, e-retailing (Capello and Ravasi 2009; Corbellini and Saviolo 2009; Storemark and Hoffmann 2012; Djelic and Ainamo 1999)
- production (manufacturing) and supplier perspective such as CMT (cut, make and trim), OEM (original equipment manufacturing and full package), ODM (original design manufacturing and full package) and OBM (original brand manufacturing and full package) production (Gereffi 1999; Bair 2005; Gereffi et al. 2001; Gereffi and Frederick 2010)
- sourcing models such as overseas buying office or third-party sourcing (Kunz and Garner 2007; Gereffi and Frederick 2010)

3.5 Sustainable Fashion Business Models

3.5.1 Incentive Green Business Model

An example of the incentive business model is that of the Slow Fashion approach where the emphasis is on retailing, consumer experience and consumption of fashion.

Slow Fashion movement

Slow Fashion is a "mental model" (Cataldi et al. 2010) which seeks a systems change for fashion production and consumption, in particular a change in how a consumer engages with fashion design. Inspired by the Slow Food movement, the Slow Fashion movement aims to reverse the approach to business growth of fashion that is cheap and rapid to mass-produced, standardized, traded in large volumes, globally ubiquitous, homogeneously served, using low-cost materials and labour and exploitative of the consumer desire for novelty by retailing new styles every few weeks (Fletcher 2010). The Slow Fashion movement encompasses a variety of approaches which include "sustainable new", "eco-fashion", "ethical" fashion, vintage and second-hand (Cataldi et al. 2010). Slow Fashion promotes small-scale production, traditional craft techniques, local materials and markets, which often results in slowing down the process of production and redistributing the power relations between the fashion creators and consumers (Fletcher 2010). A result of this systems change is a deeper awareness of the design process and its impacts on resource flows, workers, communities and ecosystems. Garments are higher priced as they take into account the true ecological and social costs, selling fewer but higher-priced items (Fletcher 2010). The ultimate aim of Slow Fashion is to develop durability in fashion which requires an understanding of how consumers connect emotionally with the garment (Fletcher 2015). Labelled as "cocreators" (Cataldi et al. 2010), a number of initiatives have developed that are collaborative consumption.

3.5.2 Life-cycle Green Business Models

Companies that have adopted a life-cycle model tend to be supply-side or manufacturing, e.g. chemical companies, fibre or yarn producers, textile factories or garment manufacturers. Teijin is an example of a company involved in recycling polyester to develop fibres through to textiles and garments supplied to retailers such as Patagonia (2013). Fashion retailers adopting this model are few, such as H&M and their "Conscious Clothing" denim range which uses up to 20 % recycled denim (H&M 2014, p. 4). An approach taken by small to micro-enterprises is zero-waste design, where the focus of their efforts is on manufacture that aims to reduce waste until there is no waste. Waste may be pre-consumer (fabric and garments left and often discarded in volume production) or post-consumer (fabric and garments no longer wanted and that may be put into textile collection bins or be discarded and destined for landfill). A number of techniques are employed, with varying degrees of waste arising from manufacture of garments.

Zero-waste design practice

This approach makes use of flat areas of fabric as in conventional fashion design and make. The whole area of the fabric is used and no waste is created. The fabric may be cut into regular shapes throughout the width of the fabric from one selvedge to the other and the garment design is arrived at through stitching the shapes together as a jigsaw or tessellation (McQuillan 2011). This method prolongs the life of a garment as individual shapes may be removed and replaced as they wear out. The zero-waste pattern cutting technique, where a minimum number of cuts are made to the fabric as it is draped, pleated and folded into the desired garment designs also leaves no waste. Some practitioners of this technique design and develop patterns to be cut for different styles on one fabric together to create two or three garments at the same time leaving no waste fabric from the cut and sew operation (Rissanen 2008, 2011).

Upcycle and remanufacture

Where the zero-waste approaches work with flat areas of fabric, the upcycle and remanufacture techniques make use of discarded clothing and textiles as well as flat fabrics (Hawley 2011). Often post-consumer but sometimes pre-consumer waste is used. While the process of design and make may be similar in the two approaches, both may leave some waste in the manufacture process and there are important differences which should be noted.

- Upcycled clothing is designed to be unique, one-off pieces of designer wear. They are often one-size and retail at higher prices in comparison with their high-street counterparts. Remanufactured clothing is designed to be manufactured to some degree of volume, there tends to be a range of sizes and comparable retail prices to their high-street counterparts.
- Upcycled designs may function as something that was not the original function (e.g. tablecloth to coat, coat to bag), whereas remanufacturing process returns the product to its original function as far as is possible (i.e. clothing return to clothing—it may not be possible or practical to return a product to exactly the same function.)

3.6 An Examination of the Remanufactured Fashion Business Model

There are few companies who practice remanufacture according to the definition posited in Chap. 1. The few who are able to are small to micro-enterprises and face many problems in scaling up their operations to a mass market volume production level. These issues will be explored in Chap. 5; however, to get a complete understanding of the processes outlined in Chap. 4, we will examine the remanu-factured fashion business model. The business model may be regarded as being composed of three basic elements: value creation, value architecture and reve-nue model (Yunus et al. 2010; Osterwalder and Pigneur 2009). Osterwalder and Pigneur (2010) extend the three elements into a business model into nine points as illustrated in Fig. 3.1.

- Value creation answers the question "what value does the business create for its stakeholders?" (Osterwalder 2004, p. 34). It assumes that an offer has no value in itself and that the offer only acquires value once it is bought and used by a target customer (Normann 2001) and that the potential for this value to be cre-ated has to be proposed to a particular type or set of customers addressed by the business activity, the product or service, and partners who create a link between the business and the customer (Dussart 2010).
- Value architecture answers the question "how is the value created and in what configuration?" (Osterwalder 2004, p. 34). This outlines the value chain for the offer, the economic agents and their roles within the value chain (Chesbrough and Rosenbloom 2002).
- The revenue model answers the question "how does a company earn money?" (Osterwalder 2004, p. 34) and requires outlining the basis and sources of income for the company and thus "determines the value and sustainability of the business specifies" (Chesbrough and Rosenbloom 2002). Osterwalder and Pigneur (2010) developed a nine-point business model canvas to examine the three basic components: value creation, value architecture and revenue model (Osterwalder 2004).

This business canvas is used to examine remanufactured fashion business model as it is described in Chap. 4 and practiced currently. As the businesses described were small to micro-enterprises, many of the issues regarding their business model will be similar to other similar sized models.

Value creation	Value architecture	Revenue model
Value proposition	Customer relationships	Cost structure
Customer segments	Distribution channels	Revenue streams
	Key activities	
	Key resources	
	Key partners	

Fig. 3.1 The business model canvas (Osterwalder and Pigneur 2010)

3.7 The Remanufactured Fashion Business Model

3.7.1 Value Creation

Value Proposition (What value does the company offer customers?)
Remanufactured fashion value proposition is qualitative in nature targeting the customer's experience and final outcome in the design. The clothing is, by the very nature of the design and product development process, unique or difficult to replicate exactly. The current sourcing and use of fabrics render it difficult to guarantee that each item manufactured will, as in conventional manufacture, be exactly like the others within a range of clothing designs. As with all fashion items, the remanufactured fashion products have symbolic meaning (Crane and Bovone 2006; Storbacka et al. 2012) when consumed and here it provides the "feel-good factor". While it may be argued that the consumer of remanufactured fashion would need to be knowledgeable about the process of making clothes and of the environmental issues to fully appreciate the products, it has been asserted by consumers and designers alike that buying of remanufactured fashion gave the same type of emotional value as buying vintage or second-hand clothing—good fun, making a statement, being seen to be buying the right things (Gam 2011). Vintage fashion has been defined as "garments and accessories which are more than twenty years old, which represent a particular fashion era, and which are valued for their uniqueness and authenticity" (McColl et al. 2013, p. 148).

Customer Segments (Who are the customers?)
Most companies engaged in the remanufacture approach have a niche market customer that is similar to that of the vintage market and are between "18 and 25 who love vintage fashion" (Ally in Dissanayake 2012, p. 125) or "between 25 and 50 years of age women who are ethically and environmentally aware and for whom fashion, style and design are important and want items that will last rather than throw away" (Han in Dissanayake 2012, p. 118). Vintage fashion consumers are largely female—though men play a role (Cassidy and Bennet 2012). The 18- to 25-year-old consumers (most receptive of fashion trends) have been regarded as propelling the demand for vintage fashion (McColl et al. 2013). While vintage fashion consumption may be viewed as indicative of consumer interest in ethical clothing as form of recycling and reusing fashion (Cassidy and Bennett 2012), an examination of the eco-fashion consumer personalities in the USA found the following three influential factors on their purchase intentions (Gam 2011):

- Fashion orientation (the desire to be well dressed),
- Shopping enjoyment and cost (need to for fun experience and low cost of buying)
- Environmental concern and eco-friendly behaviour (knowledge about the environment and desire to buy goods that were designed to lessen their impact on the environment).

3.7.2 Value Architecture

Customer Relationships (What sort of relationship does the company seek to establish with its customers?)
As most remanufacturers are micro-businesses, the type of relationship that many appear to seek is one of "push-push-pull" strategy with their consumer: making them aware of the business, the nature of the products as well as delighting them with their creativity through micro-blogging accounts such as "Twitter" (Kaplan and Haenlein 2011). Opportunities to develop personal assistance types of relationships with customers are often limited (as with most micro-businesses). This type of relationship is maybe developed either during sales, after sales or both. Micro-businesses struggle to maintain a balance between their design and manufacture for wholesale and (where there are the finances to afford) their retail business. The advantage with having a retail site (online websites or bricks and mortar stores) is the opportunity to develop this type of relationship with the customer, without which the business is reliant upon feedback (if any) from the wholesale clients or agents (Malem 2008).

Distribution Channels (How does the company distribute its products?)
Remanufacturers may distribute through wholesale to large retailers or small independent stores or they may retail through their own stores or Internet websites. As most remanufacturers are micro-businesses, it may take several years after setting up the business for them to accrue enough finances to have their own retail sites as it is not only expensive to buy/rent but also to maintain (Malem 2008) or develop fully transactional websites (Ashworth 2012).

Key Activities (Which are the activities/processes involved in the business?)
Key activities of the remanufacturer include sourcing materials, disassembling garments (where necessary), design, manufacture, marketing and distributing their products. The major emphasis for the activities is towards creating an efficient supply chain to drive down costs.

Key Resources (Which resources does the company depend on?)
The four categories of resources necessary to a business are physical, intellectual, human and financial (Osterwalder and Pigneur 2009).

- *Human resources*: people who work for the company—it may be the designer (or designers), the machine operators. As many of the remanufacturing companies are micro-sized, the designer may be the owner, manager, marketer and sales person. While the designer is central to the task of designing within the constraints of the fabric, the person responsible for cutting the patterns out of the fabric is also a very skilled person as the fabric may not arrive as a roll of fabric as in the conventional manufacture process. This cutter may or may not be the designer; however, there does need to be a good understanding of the relationship between a design, the pattern and fabric handle.
- *Intellectual*: as with many small businesses, there are limited financial resources to engage in a full copyright or branding strategy and the extent to which this is entered differs according to the individual company.

- *Physical*: all remanufacturing companies need the basic essential equipment to develop the physical product from a design idea. This may be at a shared or own studio equipped with sewing machines, cutting tables and storage areas as well as work tables. Other physical resources may be a showroom area or retail site, but again this is dependent upon the financial resources of the company.
- *Financial*: as with other small to micro-enterprises, most companies operate on revenues from retail sales and funding from bank loans, etc.

3.7.3 Revenue Model

Cost Structure (What is the cost structure like?)
Most remanufacturing companies, although they try to minimize costs, are value driven, in that they focus less on the costs and more on creating value for their products and services. Because of their small size and, or, access to finance, the current system within which to operate, there is limited potential scope to achieve either economies of scale (where costs are reduced through volume production and selling of a product) or economies of scope (where costs are reduced through incorporating other businesses which have a direct relation to the original product).

Revenue Streams (What types of revenue streams are involved?)
Remanufacturing companies have two main streams of revenue, retail and wholesale. Retail is either through online sales or physical stores. Wholesale is when the company sells the products to other retailer stores or e-retailers.

3.8 Implications for the Design and Manufacture of Remanufactured Fashion

Remanufactured fashion design is currently carried out by small to micro-fashion enterprises who produce small volumes for a niche market. Some remanufactured fashion collections have been successful through the high-street stores serving the mass market, but face difficulties due to lack of (i) sales volume to achieve sale figures/targets set by retailer and (ii) price sensitivity to the market. These commercial pressures are compounded by the lack of effective marketing strategies for the interaction between mass market (high volume, high use of current fashion trends and low price) and the remanufactured (low volume, high use of design and higher price). Ultimately, the commercial success of remanufactured fashion design is in meeting efficiencies, speed and quality issues:

- The efficiency and effectiveness of reverse supply chain.
- The optimum level of disassembly and rework needed for the recoverable garments.
- Recovery and rework cost as compared with the cost of virgin materials.

- The level of skill of labour forces for sorting, disassembly and redesign processes.
- Product strategy and the design strategy.
- Efficiency of the remanufacturing process.

New skills for the industry

The cutting operation in remanufactured fashion is, arguably, more skilled than in conventional fashion design. The cutter needs to understand not only the type of fabric in use but also pattern placement on space constraining shapes and be able to assess how the cutting will impact on the finished design. The approach to design creation is also a skill that is different to conventional fashion design as the designer is often presented with ready-formed garments that need to be reformed. These are skills that may need to be evaluated as being part of the education of a designer and perhaps new roles in the industry for design skills for manufacturing.

Networking

According to the descriptions of remanufactured fashion design processes in Chap. 4, major issues in design and manufacture are availability of technology and access to raw materials. Not all garments require disassembly prior to redesign. Some designs are developed that incorporate seams from the original garment. However, where disassembly is required, this is a time-consuming, laborious stage of the whole process and it is expensive. A consortium led by the University of Leeds and C-Tech Innovation with Madeira Threads have developed a disassembly technology using a new sewing yarn that loses its tensile strength when exposed to microwave radiation. Designers and manufacturers can choose to manufacture either whole or parts of a garment, depending on disassembly needs. The sewing yarn behaves conventionally until exposed to the radiation (Philpot et al. 2013). Although not in use commercially as yet, the speed of disassembly again suggests commercial benefits to remanufacturing. Again, Chap. 4 outlined the potential use of technology to overcome the unconventional cutting procedures in remanufactured fashion businesses. Adopting new technology involves a balance between the return on investment of using the new technology and the costs of manpower. Many businesses can ill afford new technology and their maintenance, and the remanufacturing enterprises have no exception. Possibilities may exist in developing networks or associations whereby technology could be co-owned or shared.

Access to raw materials is an issue that is a continual dilemma for the remanufacturing enterprise. The remanufacturing process described in Chap. 4 is comparable to the RemPro matrix (Sundin 2004) which demonstrates that efficiency of reverse supply chain affects the availability of remanufactured products to the customers. The apparel remanufacturer currently has little or no control over the reverse supply chain; management of the whole reverse logistic network is impeded by the cost implications, resulting in (i) high variability of quality and quantity of incoming materials and finished products; (ii) increased operational costs due to necessary additional space and labour to sort and grade and (iii) highly variable processing times that complicate production planning.

Developing collaborative networks with established textile waste collection authorities or develop product return systems is vital. Some take-back systems already exist; for instance, the Japanese polyester fibre and clothing manufacturer Teijin runs *"EcoCircle"* which network companies such as AEON and Uniqlo in Japan and Patagonia in the USA which must use their proprietary recyclable fibre to develop a range of products with various applications including garments (Sinha and Hussey 2009). In the UK, Marks and Spencer have teamed up with Oxfam with the target of reselling garments to major second-hand clothing markets. The SOEX group (global textile collectors) have developed a system called I:CO, a network of retail organizations with collection boxes for items which are returned to retailers who in turn send to SOEX for sorting and processing, a system that H&M has adopted within their retail sites. Moreover, networking with large textile collection companies can bring job creation benefits to localized markets. Textile collection companies can sort clothes that they have gathered into 140 classifications (grades). This can be expensive in terms of labour; the higher the number of grades, the greater the manpower required with associated costs. The trend is for textile collectors to no longer "fine sort" but instead to outsource this detailed sorting to partners in destination markets. The Textiles-4-Textiles consortium funded by the European Union Eco-Innovation has developed a system whereby colours and fibres such as cotton wool, polyester and blends are identified through near-infrared (NIR) technology and air jet streams automatically and quickly place the item of clothing into the appropriate receptacle (Palm et al. 2014). Although not yet in commercial use, the automatic sorting technology has the potential to not only reduce time to sort and employment costs, confidence in the type of fabric but also relocate sorting back to regions with the benefits of costs in transport and to the environment.

Quality standards

Currently, quality of remanufactured apparel is dependent on the designer and machine operator's skills and experiences, but the quality inspection process must be standardized if firms are capable of progressing from niche market volumes to mass market. Designers and manufacturers may be able to develop a quality standard for the inspection of discarded cloths (possibly through use of T4T machine) and for the final product. As quality is a key factor for the mass market, remanufacturers may explore incorporating existing final garment inspection quality standards into their process. A quality indicator would also satisfy fully the definition of remanufacture as expounded by Ijomah et al. (2007) "The process of returning a used product to at least OEM original performance specification from the customers' perspective and giving the resultant product a warranty that is at least equal to that of a newly manufactured equivalent".

National and International Trading

The confusion about the term "remanufacture" and lack of a "universally accepted definition of remanufacturing" (APSRG 2014a, p. 1) is a problem in developing legal frameworks for international trading and consumers' confidence. Issues regarding manufacturing remanufactured fashion across global trade policies require frameworks that are accepted between trading parties, e.g., the need for cleaning or fumigation certificates, the ban on intimate apparel, socks, etc., the size or weight of individual bales of second-hand clothes and the total amount imported, etc. (Hawley 2011).

As a consumer, when buying a garment, it is the brand label and the purchase receipt that are the warranties that are viewed as inducing confidence. This is the same for conventional as well as remanufactured fashion, within a local market (national as opposed to international). When developing remanufactured products across global trade policies, it is subject to trade agreements on import and export of used (core) garments. With respect to ecolabels this raises two important questions: (i) would the use of eco-labels help in international trade and (ii) are international markets necessary for the remanufactured enterprise.

As the apparel industry is customer driven, eco-labels have been regarded as an important tool in communicating and developing sustainable consumption with people, process and profit in mind. Eco-labels incorporating end-of-life management strategies such as recycling are beginning to be developed within the textiles industry. The Global Recycled Standard by Textile Exchange certifies that a firm produces recycled textiles to set criteria using post- or pre-consumer products. The label was originally developed by Control Union Certifications (CU) in 2008 and ownership was passed to Textile Exchange 1 January 2011 onwards (Textile exchange 2014). The textiles exchange website identifies up to 400 companies as having been certified (Textile exchange 2015). Within the fashion arena, the R Cert has been developed by the non-governmental body ReDress which certifies that a fashion brand has used recycled textiles in the preparation of a range of clothing (ReDress 2014).

As yet we have not identified an eco-label for remanufactured fashion. Indeed, there have been calls for a Slow Fashion label to be developed (Cataldi et al. 2010). Although global retailers have shown interest in developing eco-labels (Gallagher 2011), European retailers using textiles certified to OekoTex100, one of the three most successful eco-labels (Golden 2010), had not felt that this gave them any competitive advantage in retailing, except for babywear and children's wear where consumers were reassured about the health and chemical safety aspects of the textiles used in the designs (Austgulen et al. 2013). Those retailers felt wary of applying eco-labels on adult fashion garments because they felt that (i) their consumers either lacked knowledge or were confused about what the eco-label represented, (ii) there was a disparity between eco-labels regarding coverage and (iii) the premium placed on the final product made it uncompetitive. The decision to buy an apparel item is more complex than that for food, fridges, electronics, automotive, etc.—they found that consumers primarily looked at colour, fabric, style followed by fit and functionality and then price. As there is a growing consensus that the focus of the eco-label should be directed not at the consumer but at business to business buying based on the evidence of the success of the public and private "green procurement" policies (OECD 2005; UNEP 2007), this does raise questions about the nature of a warranty for remanufactured fashion should be.

Business and market development

Turning to the question raised about the necessity of international markets for the remanufactured enterprise, the decision to develop markets at local, national and international levels is a matter of business growth as capabilities and capacities are related to size and available finances of a company. While business growth is a decision that individual businesses need to make, there is also a need to examine this philosophically from the premise of addressing sustainable fashion consumption (Fletcher 2012, 2015; Brooke 2015). Chapter 5 will examine these within the fashion system that remanufactured fashion operates within.

3.9 Conclusion

This chapter has examined the remanufactured fashion business processes described in Chap. 4 within the contexts of other sustainable fashion design approaches. Using the business model canvas approach, the remanufactured fashion design process was examined as a business model. Implications were identified for the design and manufacture as well as the systems perspective of the fashion industry. Design and manufacture implications were related to developing efficiencies in product development (access to raw material, cutting operations and garment disassembly). Implications for the fashion industry from a systems perspective were noted to be related to networking approaches and international trade in remanufactured fashion. Issues identified were the potential for new skills education or roles within the industry, the type of warranty that should be developed for trade in remanufactured fashion and the nature of business development that would be appropriate and possible for the remanufactured business model. These issues will be examined a part of a systems perspective for remanufactured fashion in Chap. 5.

References

Ashworth CJ (2012) Marketing and organisational development in e-SMEs: understanding survival and sustainability in growth-oriented and comfort-zone pure-play enterprises in the fashion retail industry. Int Entrepreneurship Manage J 8:165–201, doi:10.1007/s11365-011-0171-6 (Springer)

APSRG (2014a) Remanufacturing: towards a resource efficient economy. The All-Party Parliamentary Sustainable Resource Group, March, http://www.policyconnect.org.uk/apsrg/research/report-remanufacturing-towards-resource-efficient-economy-0

APSRG (2014b) Triple win: the economic, social and environmental case for remanufacturing. December 2014, http://www.policyconnect.org.uk/sites/site_pc/files/.../apsrgapmg-triplewin.pdf

Austgulen MH, Stø E, Jatkar A (2013) The dualism of eco-labels in the global textile market. An integrated Indian and European perspective. Paper for OrganizaciónInternacionalAgropecuaria S.A. (OIA), 17 June 2013 http://www.oia.com.ar/documentos/ecolabels.pdf

Barney JB (1991) Firm resources and sustained competitive advantage. J Manage 17:99–120

Bair J (2005) Global capitalism and commodity chains: looking back, going forward. Competition Change 9(2):153–180 (Maney Publications, 1 June 2005)

Bellman R, Clark CE, Malcolm DG, Craft CJ, Ricciardi FM (1957) On the construction of a multi-stage, multi-person business game. Oper Res 5(4):469–503

Beltramello A, Haie-Fayle L, Pilat D (2013) Why new business models matter for green growth. OECD Green Growth Papers, 2013-01, OECD Publishing, Paris. doi:10.1787/5k97gk40v3ln-en

Brooke A (2015) Clothing poverty, the hidden world of fast fashion and second hand clothes. Zed Books Ltd., London

Capello PV, Ravasi D (2009) The variety and the evolution of business models and organizational forms in the italian fashion industry. Business history conference, http://www.thebhc.org/pub lications/BEHonline/2009/capelloandravasi.pdf

Cassidy TD, Bennett HR (2012) The rise of vintage fashion and the vintage consumer. Fashion Pract J Des Creative Process Fashion Ind 4(2)239–261, http://dx.doi.org/10.2752/1756938 12X13403765252424

Cataldi C, Dickson M, Grover C (2010) Slow fashion: tailoring a strategic approach towards sustainability. MA Thesis, Blekinge Institute of Technology, Sweden

Chesbrough H, Rosenbloom RS (2002) The role of the business model in capturing value from innovation: evidence from xerox corporation's technology spinoff companies. Ind Corp Change 11(3):529–555

Corbellini E, Saviolo S (2009) Managing fashion and luxury companies. Etas, Milan

Crane D, Bovone L (2006) Approaches to material culture: the sociology of fashion and clothing. Poetics 34:319–333

Djelic ML, Ainamo A (1999) The co-evolution of new organization forms in the fashion industry: a historical and comparative study of France, Italy and the USA. Organ Sci 10:622–37 (Sept–Oct)

Dissanayake DGK (2012) Sustainable and remanufactured fashion. Unpublished PhD Thesis, University of Manchester

Dussart C (2010) Creative cost-benefits reinvention: how to reverse commoditization hell in the age of customer capitalism. Palgrave-Macmillan, London

Dyer JH, Singh H (1998) The relational view: cooperative strategy and sources of interorganizational competitive advantage. Acad Manage Rev 23:660–679

Fletcher K (2010) Slow fashion: an invitation for systems change. Fashion Pract J Des Creative Process Fashion Ind 2(2):259–265. doi:10.2752/175693810X12774625387594

Fletcher K (2012) Durability, fashion, sustainability: the processes and practices of use. Fashion Pract J Des Creative Process Fashion Ind 4(2):221–238, http://dx.doi.org/10.2752/1756938 12X13403765252389 (Taylor & Francis Group)

Fletcher K (2015) Sustainable fashion and textiles: design journeys (2nd edn). Routledge, London

FORA (2010) Green business models in the Nordic region: a key to promote sustainable growth. FORA Green Paper, http://www.nordicinnovation.org/Global/_ Publications/Reports/2012/2012_12%20Green%20Business%20Model%20 Innovation_Conceptualisation%20next%20practice%20and%20policy_web.pdf

Gallagher V (2011) Retailers back eco-label clothing plan. 1 March 2011, http://www.drapersonline.com/news/retailers-back-eco-label-clothing-plan/5022994.article#. U94MgPldWF9

Gam HJ (2011) Are fashion-conscious consumers more likely to adopt eco-friendly clothing? J Fashion Mark Manage Int J 15(2):178–193, http://dx.doi.org/10.1108/13612021111132627

Gereffi G (1999) International trade and industrial upgrading in the apparel commodity chain. J Int Econ 48:37–70

Gereffi G, Frederick S (2010) The global apparel value Chain: trade and crisis—challenges and opportunities for developing countries. The World Bank, policy research working paper 5281

Gereffi GH, Kaplinsky R, Sturgeon TJ (2001) Introduction: globalisation, value chains and development. IDS Bulletin 32.3, Institute of Development Studies, https://www.ids.ac.uk/ files/dmfile/gereffietal323.pdf

Golden JS (ed) (2010) An overview of ecolabels and sustainability certifications in the global marketplace. Duke University, North Carolina, USA, http://center.sustainability.duke.edu/sites/default/files/documents/ecolabelsreport.pdf

Hawley J (2011) Textile recycling options: exploring what could be. In: Gwilt A, Rissanen T (Eds) Shaping sustainable fashion. Earthscan, London, pp 143–156

H&M (2014) Conscious actions sustainability report, http://sustainability.hm.com/content/dam/hm/about/documents/en/CSR/reports/Conscious%20Actions%20Sustainability%20Report%202014_en.pdf

Ijomah WL, McMahon CA, Hammond GP, Newman S (2007) Development of design for remanufacturing guidelines to support sustainable manufacturing. Robotics Comput-Integr Manufact 23:712–719 doi:10.1016/j.rcim.2007.02.017 (Elsevier)

Kaplan AM, Haenlein M (2011) The early bird catches the news: nine things you should know about micro-blogging. Bus Horiz 54:105–113 (Elsevier)

Kunz GI, Garner MB (2007) Going global—the textile and apparel industry. Fairchild Books, London

Malem W (2008) Fashion designers as business London. J Fashion Mark Manage 12(3):398–414 (Emerald Group Publishing Limited, 1361–2026)

McColl J, Canning C, McBride L, Nobbs K, Shearer L (2013) It's vintage darling! An exploration of vintage fashion retailing. J Text Inst 104(2):140–150. doi:10.1080/00405000.2012.702882

McQuillan H (2011) Zero-waste design practice: strategies and risk taking for garment design. In: Gwilt A, Rissanen T (eds) Shaping sustainable fashion. Earthscan, London, pp 83–97

Nordic Innovation (2012) Green business model innovation—policy report. Nordic Innovation, Oslo (Fashion business models)

Normann R (2001) Reframing business. Wiley, Chichester

OECD (2005) Effects of eco-labelling schemes: compilation of recent studies. Joint working party on trade and environment, OECD, JT00179584, http://www.oecd.org/officialdocuments/publicdisplaydocumentpdf/?doclanguage=en&cote=com/env/td(2004)34/final

Osterwalder A (2004) The business model ontology: a proposition in a design science approach. PhD Thesis, Universite de Lausanne, Ecole des Hautes Etudes Commerciales Osterwalder and Pigneur 2009)

Osterwalder A, Pigneur Y (2009) Business model generation, ISBN 978-2-8399-0580-0. Accessed at http://www.businessmodelgeneration.com/downloads/businessmodelgeneration_preview.pdf

Osterwalder A, Pigneur Y (2010) Business model generation: a handbook for visionaries, game changers, and challengers. Wiley, New Jersey

Palm D, Elander M, Watson D, Kiørboe N, Salmenperä H, Dahlbo H, Rubach S, Hanssen O, Gíslason S, Ingulfsvann A, Nystad Ø (2014) A Nordic textile strategy, Part II: a proposal for increased collection, sorting, reuse and recycling of textiles. Nordic Council of Ministers 2014, http://dx.doi.org/10.6027/ANP2015-513

Patagonia (2013). "Closing the loop—a report on patagonia's common threads garment recycling program." The Cleanest Line, http://www.thecleanestline.com/2009/03/closing-the-loop-a-report-on-patagonias-common-threads-garment-recycling-program.html

Philpot B, Pye A, Stevens G (2013) De-labelling branded corporate-wear for Re-use, project MPD007-007. WRAP, http://www.wear-2.com/news/De-labelling-branded-corporate-wear-Report.pdf

Porter ME (1985) Competitive advantage: creating and sustaining superior performance. Free Press, New York

Porter ME (2001) Strategy and the internet. Harvard Bus Rev 79(3):63–78

ReDress (2014) Recycled textile clothing standard informs mainstream fashion consumers. September 23, http://www.rcert.org/

Rissanen T (2008) Creating fashion without the creation of fabric waste. In: Hethorn J, Ulasewicz C (eds) Sustainable fashion: Why now? A conversation about issues, practices and possibilities. Fairchild Books, New York, pp 184–206

Rissanen T (2011) Designing endurance. In: Gwilt A, Rissanen T, (eds) Shaping sustainable fashion. Earthscan, London, pp 127–138

Sinha P, Hussey C (2009) Product labelling for improved end-of-life management: an investigation to determine the feasibility of garment labelling to enable better end-of-life management of corporate clothing. Centre for remanufacturing and reuse, Oakdene Hollins, DEFRA Clothing Roadmap Study, http://www.uniformreuse.co.uk/futureindex.html

Storbacka K, Frow P, Nenonen S, Payne A (2012) Designing business models for value co-creation. Rev Mark Res 9:51–78

Storemark K, Hoffmann J (2012) A case study on the business model of Chloé. J Glob Fashion Mark 3(1):34–41, http://dx.doi.org/10.1080/20932685.2012.10593105

Sundin E (2004) Product and process design for successful remanufacturing. PhD Thesis, Linköping's University, Sweden

Teece D (2010) Business models, business strategy and innovation. Long Range Plan 43:172–194 (Elsevier)

Textile exchange (2014) Global recycled standard version 3.0. http://textileexchange.org/sites/default/files/te_pdfs/integrity/GRS%20v3%20Draft%202.0%20-20DRAFT%20for%20Second%20Stakeholder%20Review.pdf

Textile exchange (2015) Companies certified to the global recycled standard, https://www.textileexchange.org/upload/Integrity/Standards/GRS/GRS%20Combined%20List.pdf

UNEP (2007) Background assessment and survey of existing initiatives related to eco-labelling in the African Region, http://www.unep.org/roa/docs/pdf/RegionalAssessmentReport.pdf

Warnier V, Lecocq X, Demil B (2004) Le business model: l'oublié de la stratégie?. Actes de la 13ème conférence de l'AIMS, Normandie, Vallée de Seine 2, 3 et 4 juin 2004

Wirtz BW, Pistoia A, Ullrich S, Göottel V (2015) Business models: origin, development and future research perspectives. Long Range Plan 1–19 (Elsevier)

Yunus M, Moingeon B, Lehmann OL (2010) Building social business models: lessons from the Grameen experience. Long Range Plan 43(2–3):308–325

Chapter 4
Examples and Case Studies

4.1 Introduction

The process of remanufacturing fashion clothing was analysed in five selected companies. Semi-structured interviews, field notes from direct observations and published documents were used to collect information of each of the companies and their process of remanufacturing. All the companies were found to be macro-scale independent companies, specialized in fashion remanufacturing. Both second-hand clothing (SHC) and waste textiles are used in the remanufacturing process. The following section describes the remanufacturing process practised by each of the companies.

4.2 Overview of the Fashion Remanufacturing Process

4.2.1 Company A

Company A was founded in 1997 by two fashion designers. Those designers first remade their clothes to wear for going out to clubs in the early 1990s. The compliments they received for the designs encouraged them to set up a small business to deconstruct second-hand clothes and redesign them into twisted tailored garments. The company is now established as a clothing label, a shop and a small design team specializing in fashion remanufacturing. The target customer group is women/men who really have a strong identity, love to be unique and prefer the individual look.

The materials used for remanufactured fashion are either recycled, fair trade or organic. Those fabrics are mixed with best quality SHC such as men's suits, T-shirts, knitwear, silk smoking jackets and coats made from technofabrics,

© Springer Science+Business Media Singapore 2016
P. Sinha et al., *Remanufactured Fashion*, Environmental Footprints
and Eco-design of Products and Processes, DOI 10.1007/978-981-10-0297-7_4

vintage ties, dead stocks, denims and dresses. SHC is obtained from charity shops and SHC collection/sorting centres. Suiting materials are used as the primary material for the remanufacturing process. Collected men's suits are checked and dry-cleaned before being reused. Other SHC and the fabrics are also cleaned, if necessary, and sorted according to the colours or fabric type.

The process of remanufacturing is fairly different to the manufacturing process of an original fashion product. Designers attempt to transform best quality second-hand garments into timeless, unique fashion pieces. Fashion collections do not conform to fashion trends because the purpose is to create a sustainable fashion collection that represents a unique theme. "…we don't really follow trends. We kind of do what we want to do and we feel, obviously…but the collection has a definite style, a definite mood, a definite movement…", says the business manager.

With dimensionally restricted materials, designers follow a very creative and practical approach to explore design ideas. For example, a trouser or a coat is experimented upside down on the mannequin, suits or jackets are mixed with ties to form a dress, or a shirt is draped on the mannequin to form a skirt (Fig. 4.1). Draping technique is largely being used, and the garments are disassembled only when needed. If there is a requirement to disassemble some of the component parts, that is being carried out simultaneously while exploring the design ideas. Designers' attempt is to keep the disassembly minimal as it is a time-consuming, manual operation. Many garments are created to be multistyled, i.e. one garment can be worn in many different ways.

Fig. 4.1 Draping a shirt to form a skirt (*source* Authors)

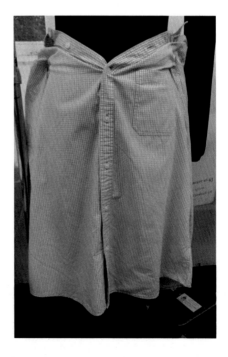

The company takes part in several catwalk events and trade shows annually and receives short-orders or rolling orders from the buyers nationally and internationally. The product design seems to be standardized throughout the order; however, the fabrics are non-standard in terms of colours and material type. Therefore, repeatability is possible through the design, yet each garment differs by colours or fabrics. In addition to the design range, company A creates one-off pieces to sell in its own shop. No two garments are same, even though they are cut from the same patterns, because the fabrics or the colours are always different from each other.

Selling remanufactured fashion has been successful through the high-street stores serving the mass market; however, there are difficulties to continue due to lack of sales volume to achieve sales targets set by the retailer and the lack of price sensitivity to the market. Nevertheless, the collection is non-standard and the remanufacturing process is highly labour-intensive, resulting in a high price, which is a barrier to enter the mass market.

4.2.2   Company B

Company B is a fashion remanufacturing business and also a social enterprise. The aim of the business was to explore textile recycling possibilities and local community development. According to the designer, her work on recycling clothes started twenty years ago when she was doing a project with Oxfam to investigate the possibilities of reusing discarded, donated garments. Meanwhile, by looking at the amount of waste produced due to student sampling processes or the concept of fast fashion, the designer realized that there was a gap in understanding what sustainable fashion meant. Therefore, she had set up the social enterprise to experiment different techniques as a designer to utilize the waste as a resource and also to teach design skills to local people.

Company B uses discarded clothes to produce collections of sustainable designs. The enterprise community recycling workshops are conducted to teach participants the skills of remaking and extending life of the garments in their own wardrobe. Designer believes that knowledge transferred through community workshops helps to reduce air miles associated with clothing and improve local design and manufacturing capabilities. The business targets women in the local community, who loved recycled products and express an interest to learn fashion designing and sewing skills.

The designer uses discarded apparel, household textiles (curtain, duvet covers, etc.) and waste fabrics from textile factories as raw materials to the process. According to the designer, the process cannot be started with design sketches as the final design is largely governed by the existing material stock. Therefore, the approach is to break the ground rules of fashion designing and enjoy the freedom of creativity. Initial design ideas are explored by draping the fabrics, creating different shapes and making toiles.

Fig. 4.2 Remanufactured
fashion (*source* Authors)

Most of the garments are fully disassembled by cutting along the seams using pair of scissors, in order to get a sufficient length and width of the material to work with. However, when there are complex design and construction features included in discarded garments, the designer tends to maintain those features in the new design without destroying them by disassembling. Designs are mainly one-off pieces, and it is challenging to produce few repeats due to fabric restrictions. Figure 4.2 shows some of the sample garments made by the designer.

There are community workshops held by the designer for the local community to enhance their design and stitching capabilities. The participants create few designs for the company and also for their own use. Those garments are sold through ethical or sustainable fashion shops, through market stalls or through personal channels.

4.2.3 Company C

Company C is operated by a designer who is inspired by couture styling. Company C started in the year 2008 by the designer as she wanted to promote sustainability in fashion. Discarded denim fabrics are collected and transformed into high-quality, unique design pieces. Denim fabrics are used as the main raw material due to various reasons. Cotton is used to make denim fabrics, yet cotton is an unsustainable fibre in terms of water and chemical use, and therefore, reusing denim fabrics saves the resources and environment. Moreover, denims never go out of fashion; the more you wash and use it, the more features it gives.

The main source of denim fabrics are denim trousers, collected from charity shops or consumers. Those trousers are completely disassembled before remanufacturing them into new designs. Company employees design students or people from the local community as part-time staff to disassemble denim trousers. The process of disassembling is manual and time-consuming, and quality of disassembly could be a problem as many of the employees do not follow the instructions provided. Figure 4.3 shows some of the denim trousers which are disassembled.

Due to the narrow width of the trouser panels, redesigning process is challenging, and therefore, the creativity of the designer plays an important role in the design process. In order to produce an attractive, unique design, disassembled pieces are experimented on the mannequin by creating various shapes and styles. Each of the experiment is photographed and analysed later to decide the best possible combination of shapes and pieces. Successful shapes are selected and proceed into a final design, as shown in Fig. 4.4.

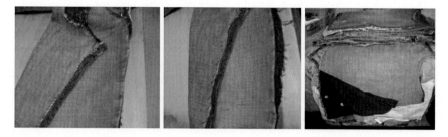

Fig. 4.3   Disassembled denim trousers (*source* Authors)

Fig. 4.4   A dress made from discarded denims (*source* Authors)

The company targets high-end consumer with expensive pricing, because the garments are individually designed and manufactured. Moreover, the whole process of remanufacturing involves more time and skilled labour, compared with the traditional manufacturing process. The company attempted to sell the products through existing retailing channels and however compelled to pull out due to high price. Nevertheless, the price should reflect the amount of time and effort spent on making that fashion, because it may take 2–3 days or even more to complete one design. Those designer-made, expensive pieces would be a way to promote slow fashion among eco-minded consumers.

4.2.4 Company D

Company D is a forward thinking women's wear vintage clothing and accessory label that focuses on giving a new life for tired vintage clothing. The business, founded in the year 2009, was initially supported by John Moore's enterprise programme and the Prince's Trust funding. The company is owned and managed by the designer herself. She was trained in adult education and used to coach people of all ages to develop their skills and creativity. She teaches fashion and textiles in school, in colleges and in the community. With the help of the School for Social Entrepreneurs programme, she established "Work and do" workshops to bring communities together and promote to create something new from old.

The company sources SHC from a wholesaler who deals in vintage clothing. Previously, company made an attempt to collect discarded clothing from the consumers and charity shops, but it was a time-consuming task and the designer was unable to collect sufficient amount of materials to fulfil the requirements. Sourcing from a wholesaler was found to be the best option where the designer can purchase only what she needs. As the target customer group is the girls within 18–25 age group, discarded, large-sized garments with bright colours and floral prints are ordered from the wholesaler. An order contains around 40 kg of clothing and delivered based on a request made by company D. Due to the long relationship the company maintains with the wholesaler, the wholesaler has a better understanding of the fabrics and prints required by the company and supplies accordingly. Those pieces of vintage clothing are mixed with remnant fabrics purchased from other merchants, in order to produce sustainable fashion.

The disassembly operation could be either complete unpicking of a garment along the seams (if the garments were small in size, e.g. size 8–14), or using a pair of scissors to cut along the seams if the garment is large enough to obtain sufficient fabric for the intended design. Since the designer worked mostly with plus size discarded garments, she could cut the new pattern out even without disassembling the whole garment. "Sometimes I don't disassemble them, that's why it takes quicker time", says the designer. Instead, a dress could be opened up by cutting along the waist line and from the side seam. She found this as an easy and quick method when compared with the effort and time taken in the disassembly process.

The designer attempts to create fairly simple, trans-seasonal designs in three sizes: small, medium and large. Most of the dresses are semi-fitted, and the waist band is elasticated to minimize the need for sizing. For example, as shown in Fig. 4.5, a "small"-sized dress could be worn by a size 8–12 or even a size 14 person; this reduces the wastage and fit issues.

Moreover, the designer experiments with appliqué and screen printing techniques in remanufacturing new from old clothes, as shown in Fig. 4.6. The designer explores the possibility of using screen printing technique to give a fresh look to

Fig. 4.5 Multisized garments (*source* Authors)

Fig. 4.6 Appliqués and screen printing techniques (*source* Authors)

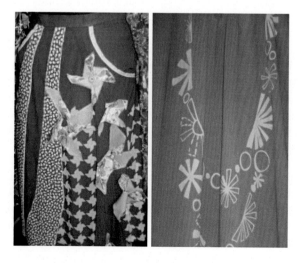

tired vintage material; however, there are several challenges such as the additional cost that need to be incurred and the increase of the price of screen-printed clothing.

The remanufactured garments are sold through vintage shops and market stalls. Those fashions are proved to be attractive to the youth, mainly because vintage has become a fashion now, as the designer experienced. However, selling those remanufactured fashions through existing sales channels is difficult due to the high price of those designer-made pieces.

4.2.5 Company E

Company E, located in London, is founded in 2006 by a designer and her business partner. It operates as a small enterprise, comprising four full-time employees and few fashion students spending their internships. According to the designer, her interest in ethical fashion began while studying for the undergraduate degree in fashion design where she created a fashion collection entirely from organic and recycled materials. While doing the collection, she came to know the amount of good-quality clothes that are thrown away by the consumers and realized the possibility of creating new clothes from recycled fabrics.

The company designs and produces innovative, quality women's apparel and accessories which are made from hand-picked, locally sourced, and recycled fabrics. Company's own brand continues to grow and wins a number of awards including Trefor Campell Award for Creative Enterprise and SME (Small Medium Enterprise) Innovation Award.

Company collects discarded clothing from charity shops and commercial recycling companies or through consumer donations. Those discarded pieces of clothing are disassembled and mixed with remnant fabrics to remanufacture them as best quality, sustainable fashion collection. The main reason for using remnant fabrics is to produce repeats and volumes. Initial design ideas are generated by mixing various materials and colours to explore the possibilities of creating a commercially viable product.

Design finalization is largely influenced by the quality and quantity of available fabric stock. If the quantity of the material is restricted, yet the fabric is in good quality, the designer tends to create one-off piece for the high-end market. If there are sufficient quantity of materials available in one particular fabric type and colour, the final design aims at producing volumes. To enhance the capability of producing volumes, company also collects fabrics with similar types and prints over a period of time.

Disassembly is not a compulsory operation for the designers in this process. Production pieces are directly cut from the discarded apparel without disassembling the entire garment, in order to save time and labour cost. Moreover, knitted fabrics are preferred than woven fabrics when dealing with dimensionally restricted fabrics, because of the fact that knitted fabrics could be stretched to adjust to the production pattern, if there is a little shortage.

There are certain set of rules followed by the pattern cutting operator when cutting the panels for bulk orders, and the designer highlighted a new skill set required for the pattern cutting operator, in order to manage the process efficiently and effectively. Operator must possess design skills and technical ability in order to make independent decisions during cutting as it is not a straightforward process. Basically, a design has standard pattern and standard set of rules to cut by, yet the fabric is not standard throughout the order. For example, when there is a panel T-shirt being produced, same production patterns will be used throughout the production order, but the fabrics and colours of the panels could be different from one garment to the other. However, there are certain rules to cut them out so that they all look basically the same design.

Company has been producing fashion collections to exhibit in London Fashion Week since 2009. The collections had been displayed as a part of Estethica, the ethical arm of the British Fashion Council. There are two catwalk events organized every year: spring/summer and autumn/winter. As a result of catwalk events, company receives production orders from fashion retailers nationally and internationally. The order quantities ranged from 70 to 100 pieces for the selected styles from the sample collection. However, as a sustainable fashion label, company recently decides not to operate as an ordinary fashion brand that shows two collections per year. Instead, the company issues a trans-seasonal lookbook with a colour story that includes previous designs and few new designs. The idea is to obtain bulk orders for those trans-seasonal products and not to restrict to the seasonal time boundaries. "…we will offer that as whole sale, trans-seasonal so that people don't have to get their order by certain date, get it to the shop by certain time to fit with all their other stock", says the designer.

One-off pieces are manufactured in-house, and the small orders were outsourced to several manufacturing facilities. Moreover, company E developed manufacturing capacity in Bulgaria with the aim of mainstreaming remanufacturing fashion business and planning production to bulk orders in that manufacturing facility.

Company E sells its products online as well as through several retailers nationally and internationally. Company has also teamed up with high-street retailers to produce remanufactured fashion collections.

4.3 Implications for Mass Manufacturing

The case studies reveal that the fashion remanufacturing business currently operates in niche market level, and the industry has yet to develop the process into a more mainstream business model. Reducing the environmental burden caused by waste textiles would presumably be possible through remanufacturing greater volumes, potentially through mass markets. The following implications are raised through the case studies:

4.3.1 Process Input

In a conventional manufacturing process, fabrics are purchased according to the finalized designs, whereas in remanufacturing, the acquisition of fabrics is the first stage, and the design process is largely governed by the characteristics of SHC. Producing volumes from a particular design is complicated due to fabric restrictions, and the designers have to challenge the traditional production norm of creating volumes. Conventional mass production divides an order quantity into different sizes and colours, and each garment must be produced from one fabric type. Obtaining large quantities of fabric from one particular type is difficult in a remanufacturing process, and therefore, production volumes in remanufacturing mean standard design throughout the production order, yet different types of fabrics and colours could be experienced from one garment to the other.

4.3.2 Cutting Operation

The cutting operation in remanufacturing is manual, with each garment being hand-cut individually, and therefore, cutting cost is higher than the conventional manufacturing process. In the mass manufacturing process, fabrics are purchased in bulk and several garments are cut at once by laying several fabric plies and using modern cutting technologies. Obtaining several plies from regular shapes is difficult in the remanufacturing process due to high variability in fabric shapes recovered from SHC. However, a technology similar to that used by leather cutting machines, combined with a pattern-making software, could be a possibility to increase efficiency in creating volumes. Leather cutting machines allow cutting required shapes over an irregular-shaped single ply and to make timely corrections to the patterns. By using an inbuilt projector camera, the user can place the digital patterns effectively in an irregular-shaped material and adjust the cutting direction in accordance with the material. This kind of a technology would help to develop an effective cutting operation in fashion remanufacturing which would increase the output of finished products and reduce material loss and eventually increase the effective utilization of the materials.

4.3.3 Garment Assembly

Garment assembly in the remanufacturing process is fairly standard as in the conventional process and is carried out using industrial sewing machines. However, production is based on an individual/whole garment system, i.e. one operator assembles the entire garment. This is an effective system when a large variety of garments have to be produced in extremely small quantities. The widely used system for mass production in the industry is the progressive bundle system, where the entire production is broken down into sub-operations, and the garments are gradually assembled as bundles of cut panels moved through operations. Each operator is responsible to perform one sub-operation (e.g. sewing the hem, sewing the centre back seam). However, this system may not be suitable for mass production in remanufacturing due to frequent variations of colours and fabric types. Given that the design remains unchanged, operators can still be allocated to perform specific sub-operations for the entire production order. However, line efficiency could be decreased because the thread colour needs to be changed when the fabric colour changes, and the machine needs to be adjusted depending on the fabric type. Nevertheless, adopting this manufacturing system would be possible if the operators are provided with extra thread cones, bobbins are filled up with the required thread colours, and if the operator is capable of adjusting the machine according to each fabric type.

The modular manufacturing system can be considered as the most suitable system for fashion remanufacturing, given the high degree of style changes. The modular manufacturing system consists of small groups of highly cross-trained operators organized into modules with a high operator empowerment to make decisions, which is best suited to the circumstances of remanufacturing. A work module can be created to produce the entire garment or to produce sub-assembly units. Manufacturing layout is U-shaped with more machines than operators. These excess machines with special features and attachments are required to handle style variations in remanufacturing. Operators usually work in a standing position and are interchangeable between tasks. The production unit can be one garment or small bundles. This system is characterized by greater flexibility, better quality control and short throughput times.

4.3.4 Quality Standards

It was noticed through the case studies that no quality standard is implemented to test the final quality of the products. Quality is controlled throughout the production process only by designers' experience and knowledge. However, in a mass production environment, the manufacturing operation is performed by a number of machine operators, and it is required to have a quality standard in order to maintain a consistent quality level throughout the order. In a modular manufacturing system, achieving the required quality level is the responsibility of the operator. The idea of satisfying the customer is maintained at each work station, and every operator is considered the customer of the previous operator. In this way, the quality standard is maintained within the production line, thus avoiding the time and labour requirement for final quality checking of the finished garment.

4.3.5 *Distribution and Retailing*

As per the evidence from the case studies, remanufactured fashion still fails to reach the high street mainly due to two reasons: one is the high selling price and the other is that the design studios can only standardise a design but not the fabric as the access to volumes of the same fabric are constrained by the system that the designers currently work within. Company A used to sell its products to a major high-street retailer, but was compelled to pull out because the fashion collection appeared non-standardized, and also the products were not price-sensitive to the market. This challenge was overcome by company E by collaborating with a mass market retailer. However, the collections are sold only at retailer's online shop because the volumes from one design are not in sufficient quantities to deliver and sell in the retail shops around the UK. Most of the fashion retailers are still not prepared to take the risk of having non-standard fashion collections in store, yet the extent of growth of this business is dependent on the commitment and involvement of larger retailers. Remanufacturing firms are currently attempting to build up relationships with large retailers because creating ranges for mainstream retailers is the best way to approach the mass market.

The study showed that the price of a remanufactured garment may be high mainly due to the time spent redesigning the garment and that bulk production does not take place (which would normally achieve economies of scale). Company D and Company E attempt to produce multiples of standard simple designs to minimize the time spent in the design stage which in turn reduces the cost. This must be way forward in approaching the mass market, rather than creating complex and unique products at high prices. Nevertheless, cheap price encourages consumers to buy more and dispose garments frequently, which supports a culture of fast fashion. The best option may be to make the consumer aware of sustainable fashion and the environmental benefits of their purchasing decisions, so that a consumer may make a sustainable choice even at a higher price. However, slowing down the consumption rate is less preferred by the retailers because their sales and profits are boosted by the fast-changing fashion cycles.

4.4 Conclusion

The process of fashion remanufacturing challenges the typical fashion design process that starts with the design sketch. Creation of a remanufactured collection begins with the analysis of materials obtained from discarded garments and exploring the ways of redesigning and reconstructing them, by creating various shapes. The method used to remanufacture is very much depending upon the size and the usable material space of the discarded item. Therefore, designers' creativity is a key factor for the successful completion of the process that creates timeless fashion. The real challenge for the designers is to explore new avenues in sustainable fashion design by superseding the rules in the conventional fashion design process.

Chapter 5
Systems Requirements for Remanufactured Fashion as an Industry

5.1 Introduction: A Comparison Between Conventional and Remanufactured Fashion Design Processes

Chapter 1 identified that remanufacture has been noted as a vital element in the circular economy (ASPRG 2014a, 2014b; Savaskan et al. 2004; Sundin 2004; Guide et al. 2003). It has been proposed that rather than perceive waste management as discrete set of operations (of generation, collection and disposal), it should be viewed as a system where waste is regarded as "part of a production system, the relationship of waste to other parts of the system is revealed and thus the potential for greater sustainability of the operation increases" (Seadon 2010). To that end, this chapter examines the issues that confront the development of an efficient fashion remanufacturing process. To comprehend the issues faced in remanufactured fashion, it is important to understand the conventional fashion design process practiced in companies of all sizes. Chapter 4 outlined case studies that described the processes of a number of fashion designers involved in upcycling and remanufacturing. The conventional and remanufactured fashion design processes are illustrated in Figs. 5.1 and 5.2. The design process comprises a sequence of activities which occur in a logical order, from concept formation to the final product. Although in real time it is difficult to state exactly where one set of activities ends and another starts, they are described as phased for ease of examination. Following the classic five-phase sequence, the differences in the phases and activities are presented in Table 5.1. The main points of difference are as follows:

- *Raw materials used*: virgin fabrics in conventional and second-hand clothing (SHC) in remanufactured.
- *Design trend research conducted*: colour, fabric and styles for conventional styling in remanufactured as fabric and colour are as found in SHC.
- *Range strategy to be developed as samples*: confirmed prior to concept development in conventional process, this is delayed and can only happen when samples are being developed due to constraints placed by pre-cut fabric shapes.

© Springer Science+Business Media Singapore 2016
P. Sinha et al., *Remanufactured Fashion*, Environmental Footprints
and Eco-design of Products and Processes, DOI 10.1007/978-981-10-0297-7_5

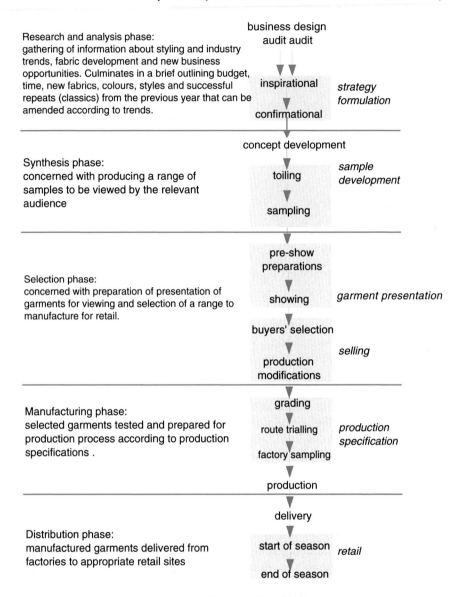

Fig. 5.1 A general fashion design process (based on Sinha 2000)

- *Sample development*: shapes cut on roll of fabric in conventional process, is restricted and decisions need to be made regarding disassembly for remanufactured.
- *Standardization of designs*: can be for colour, fabric and styles in conventional, only style in remanufactured (colour and fabric restricted by what is available).
- *Volume production*: can be in hundreds of any unit for conventional, restricted to fabric availability in remanufactured process.

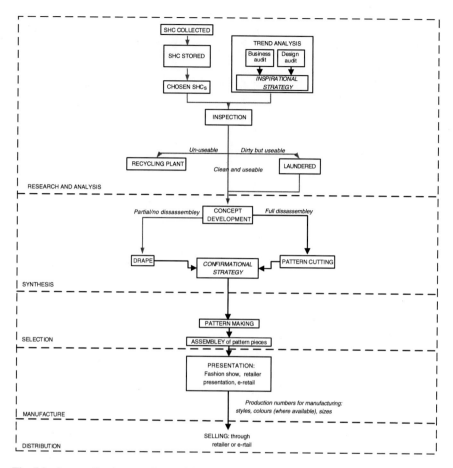

Fig. 5.2   A generalized remanufactured fashion process (based on Dissanayake 2012)

5.2 Remanufactured Fashion as a Reverse Supply System

A supply chain has been described as consisting of all the activities involved in delivering a product from raw material through to the customer including sourcing raw materials and parts, manufacturing and assembly, warehousing and inventory tracking, order entry and order management, distribution across all channels, delivery to the customer, and the information systems necessary to monitor all of these activities (Quinn 1997). The fashion design process is an element within the supply chain and as may be seen, the activities and stages in remanufactured fashion are as those of the conventional fashion design process but with the addition of collection of SHC, cleaning, sorting and storage, disassembly, prior to remanufacturing, distribution and selling. This is similar to the main activities in the reverse supply chain which include gathering used products, classifying and segregating

Table 5.1 Comparison between conventional and remanufactured fashion design process (*Source* Authors)

Is specific to conventional	Applies to both	Is specific to remanufactured
Research and analysis phase		
The larger a company, the greater volume production is required; the more risk averse they are and a greater the amount of business and design trends, research and analysis is conducted to promote confidence in new ideas Inspiration is through the research conducted but must be applied to current product ranges through creating mood boards and colour boards, to generate a theme for the design directions (Studd 2002; McKelvey and Munslow 2003) All companies will produce repeat designs to leverage previous season's successful designs; however, the ratio of repeat:new design may be as high as 70:30 in a very large mass market high volume retailer and the opposite for the design company (Sinha 2000) New fabric colours and styling will be outlined. Fabric will be sourced and ordered on the roll	Information is collected in general market trends, target market consumer needs, preferred product characteristics, forthcoming trends on colours, fabrics, silhouette, trims and design details (Sinha 2000; Burns and Bryant 2002; Rosenau and Wilson 2006; Keiser and Garner 2008) The smaller companies place greater emphasis on individuality, conduct less market and trend research but increase personal inspirations and experimentations	The designs are constrained by the colours and types of fabrics (second-hand clothing) available. Therefore, initial design strategies are limited to developing a range of styles which will be experimented upon The equivalent of fabric research for the remanufactured fashion designer is in obtaining second-hand clothing and textiles. This phase is extended for the remanufactured design process in collecting and working with the second-hand clothing collected *Collection of SHC (second-hand clothing)* Currently mainly carried out by the remanufacturing firms themselves. In order to have a better control over input materials, it is vital that remanufacturing firms makes relationships with textile waste collection and sorting companies *Sorting of SHC (second-hand clothing)* Currently manual and therefore labour intensive and expensive. The SME designers tend to sort the SHC into colours and, due to limited storage space, they do not store very much *Quality tests* Currently there is no testing of the fabrics other than an inspection by eye of the designer to decide whether the fabric is appropriate for the design *Cleaning* Not carried out at every design firm, depends on the designer's thoughts about material, any stain, etc.—testing individual garments for stain marks is time consuming and labour intensive

(continued)

Table 5.1 (continued)

Is specific to conventional	Applies to both	Is specific to remanufactured
Synthesis phase		*Disassembly*
Garment sketches developed into technical drawings (either manually or through CAD)		This is highly labour intensive and time consuming, complicated and difficult to standardize since every garment is different. The degree of disassembly is dependent on the design of the new garment—attempts to minimize the time spent on disassembly include: incorporating existing features and seams into the new design, cutting along the seams rather than unpicking the stitches, and cutting new patterns without disassembling the garment when there is a sufficient fabric space
Technical drawings are reviewed and selected for pattern creation/development		
Prototype samples for each design are made and tried on for fit, appearance, measurement and overall performance (Burns and Bryant 2002; McKelvey and Munslow 2003)		
Pattern creation	*Drape*	*Redesign*
Patterns may be created by hand or though software	Some designers prefer to drape, working with the fabric on mannequin to create the design and then transfer it into a two-dimensional pattern piece instead of sketching	The design process is largely governed by the characteristics of textiles or clothing obtained. Obtaining large quantities of one particular type fabric is difficult for remanufacturing firms and therefore, when producing in volumes the design may be standard, but fabric type and colour may differ from one garment to the other
The patterns created at this stage may or may not be production patterns—if the garments are sold through retailers, they may need to be amended according to the retailer's specifications	This enables the designer a great deal of creativity and freedom to explore three-dimensional shapes with actual fabric; however, this process consumes more time and fabrics than the former (Burns and Bryant 2002)	*Pattern creation*
		Production patterns for every garment are created manually by the designers—a very time-consuming and labour-intensive process because every garment is different and less repeatable
Selection phase		
The samples can viewed in by retailers also as internal "range meeting" with the selection committee	Company prepares a sample range of designs to show at either an industry fashion show, in-house fashion show or show rooms	
	All companies need to outline style numbers, colours, sizes and quantities for bulk manufacturing (Sinha 2000; Burns and Bryant 2002)	

(continued)

Table 5.1 (continued)

Is specific to conventional	Applies to both	Is specific to remanufactured
Manufacture phase		
Retailer sourcing decisions on whether the production will be domestic or offshore are determined by the factors such as manufacturing capacities, cost, quality control standards, expected turnaround time (Burns and Bryant 2002; Kunz and Garner 2007)	Bulk manufacturing for the selected styles requires information such as production numbers, quality standards, and size ranges from the buyer	*Cloth cutting* Due to the various shapes and dimensions of the materials, it appears to be difficult to use more than one ply in the cutting operation, even for reparable styles. This can slow down the garment assembly process. One SME employed more cutters than would normally be the case for this operation
Retailer specifications Amendments to the design are carried out as required from the retailer		Cutters need to be skilled and experienced: sometimes they may have to make slight adjustments to the production pattern depending on the fabric restrictions
Cloth cutting Is mechanised and fabric is cut in several plies to shorten the manufacture time Patterns are cut exactly as set out in the lay plans		Therefore, the cutting operator needs to be knowledgeable not only about the cutting operation, but also the other aspects of garment manufacturing such as pattern creation, design and assembly operations—to ensure maximum efficiency in use of fabric
Garment assembly In mass production, the progressive bundle system is the norm, where the entire production is broken down to suboperations; bundles of cut panels move through operations to gradually assemble the garments		*Garment assembly* Production based on individual/whole garment system, i.e. one operator assembles the entire garment, which is an effective system for large varieties of garments produced in extremely small quantities
		Volume production implications are a decrease in line efficiency as frequent changes may be needed, e.g. thread colour, sewing tension and needles/needle plates, thus slowing down the manufacture speed

(continued)

Table 5.1 (continued)

Is specific to conventional	Applies to both	Is specific to remanufactured
Distribution phase		
	Manufactured orders are delivered to the retail outlets for selling to the public	Some SME fashion designers have sold through the high street shops, e.g. Junky Styling (Top Shop) and Goodone (Tesco), but the most reliable method as yet has been sustainable fashion boutiques and online retailing due to lower volumes
	Once garments are sold, retailers collect and analyse the weekly, monthly and seasonal sales of the products	
	Manufacturers use the data from their orders for manufacture from the buyers	Remanufactured fashion cannot guarantee standardized collections or price sensitivity to the market. Additionally, only the design is standardized, but the fabrics are different from one garment to the other and online customers have to understand that they might not get the design with exactly the same fabric
	This information is used in the research and analysis phase for the design of the next season's collection	

Fig. 5.3 A simplistic representation of a reverse supply chain system (*Source* Sinha et al. 2014)

them and transporting them to the appropriate locations for reuse, remanufacture or recycling (Majumder and Groenevelt 2001). A reverse supply chain has been defined as "the movement of products from the customer back to the seller or manufacturer… the reverse of the traditional supply chain movement of products from seller to customer" (Kahhat and Navia 2013). It has been argued that every supply chain has forward and reverse logistics (Veiga 2013). A supply chain will also encounter reverse logistics activities, through returns, repairs, reuse and recycling. Products have the opportunity to return to the same supply chain in their reusable form as depicted in Fig. 5.3.

The forward flow of a supply chain is scheduled and processed by manufacturers and retailers within a certain time frame, whereas the reverse flow is initiated by the consumer. This makes managing an efficient reverse supply chain a challenge in terms of capacity planning, controlling and gaining profit from recovery activities and requiring additional consideration in planning, designing and the control of its activities. The situation could be further complicated because the customer may return the product during the product's life cycle, at the end of use and at the end of life, and each type of return requires an appropriate reverse supply chain to optimize value recovery (Fleischmann et al. 1997; Michaud and Llerena 2006; Nasr and Thurston 2006).

Chapter 2 discussed reverse logistics and it can be defined as "the process of planning, implementing and controlling the efficient and effective inbound flow and storage of secondary goods and related information for the purpose of recovering value or proper disposal" (Kumar and Chatterjee 2011). It has been observed that successful reverse logistics solutions could only occur by efficiently integrating forward and reverse flows (Krikke et al. 2001; Stock 2001). The key objectives of reverse supply chains are product acquisition, reverse logistics, inspection and disposition, remanufacturing and marketing (Kumar and Chatterjee 2011).

Remanufactured fashion design is currently carried out by SME fashion designers who produce small volumes for a niche market. Greater environmental impact would presumably be possible through greater volumes, potentially through retailing in mass market volumes. While some remanufactured fashions have been successful in through the high street stores serving the mass market, there are difficulties due to lack of sales volume to achieve sale figures targets set by retailer and price sensitivity to the market. These commercial pressures are compounded by the lack of effective marketing strategies for the interaction between mass market (high volume, high use of current fashion trends, low price) and the remanufactured (low volume, high use of design, higher price). Ultimately the commercial success of remanufactured fashion design is in meeting scale, speed and quality issues, which are impacted by:

- The efficiency and effectiveness of reverse supply chain—how the reverse supply chain is managed.
- The optimum level of disassembly and rework needed for the recoverable garments—a cost/benefit analysis between disassembly:design.
- Pricing strategy to compare favourably with those products made with virgin materials.
- A skilled labour force in the sorting, disassembly and redesign processes.
- Product and design strategy that enables mass market production volumes.

The efficiency and effectiveness of reverse supply chain—how the reverse supply chain is managed

The efficiency of reverse supply chain affects the availability of remanufactured products to the customers (Guide and Van Wassenhove 2001; Guide et al. 2003). The apparel remanufacturer currently has little or no control over the reverse supply chain, and the management of the whole reverse logistic network is impeded by the cost implications, resulting in:

(i) high variability of quality and quantity of incoming materials and finished products;
(ii) increased operational costs due to necessary additional space and labour to sort and grade; and
(iii) highly variable processing times that complicate production planning.

Technologies to help efficiencies in collection, sorting, disassembly and manufacture, as discussed in Chap. 4, are usually prohibitive in terms of costs and out of scope for enterprises that currently engage in upcycling or remanufacture as they tend to be small to micro enterprises. Relationships between the various supply chain stakeholders (forward as well as reverse) are therefore important to develop.

5.3 The Current Fashion System

Systems have been described as "a set of things—people, cells, molecules or whatever—interconnected in such a way that they produce their own pattern of behaviour over time" (Meadows 2008, p. 3). The General Systems Theory (GST) was first introduced as a theoretical framework for the study of biological development philosophies (Bertalanffy 1950) and became applied in the study of many fields including organizations, management, engineering, economics and system dynamics, because it enabled the study of individual behaviour within the environmental contexts (Forrester 1976). The fashion system may be regarded as being a set of subsystems (Fig. 5.4) which influence the ebb and flow of the primary

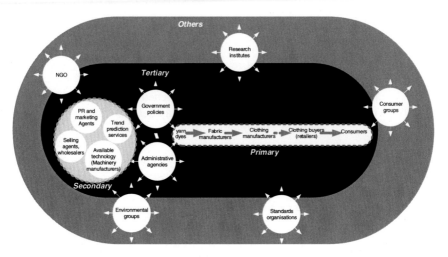

Fig. 5.4 The fashion industry subsystems (*Source* Authors)

activity of the fashion industry—that of manufacture and retailing of clothing and
fashion items. The subsystems may be described as:

- Primary (direct) subsystem is the decision-makers in the production/consump-
  tion context (manufacturers, retailers, consumers).
- Secondary (indirect) subsystems influence the decision-making of primary subsystem through their decisions (e.g. trend forecasters influence the large fabric manufacturers regarding the colouration of the fabrics and thus the choices available to
  garment manufacturers, or technology available for the manufacture of garments).
- Tertiary subsystem is the government and administrative groups who set the
  legal, political and economic framework within which the industry members
  operate.
- Other subsystems are groups or organizations in which, although not directly
  connected to the fashion industry, they try to influence the behaviour of all
  industry members to improve the status quo.

The individual parts of the systems (elements) are interconnected with either physical flows or information that together form a system (Meadows 2008). Identifying
and using beneficial relationships and linkages between different parts of a system
is key to optimizing the whole (Charnley et al. 2011; Mingers et al. 2010).

Taking a systems thinking approach (Meadows 2008), it may be observed that
the current fashion system is important with several functions:

- From the individual consumers' perspectives, the fashion system's function is
  cultural and a tool for self-expression. Hamilton (1987) proposed the concept
  of "dress" as a cultural subsystem, based on technological, social structural
  and ideological components. Kaiser et al. (1995) coined the term "appearance
  modifying commodities" to define the functions of clothing and fashion in communicating and constructing personal identities; this has economic value in the

market place as it affects buying behaviour. Fashion consumption is concentrated within the developed nations where 72 % of global imports of apparel were into the European Union, the USA and Japan and growth in regions such as Canada, Korea and Australia (WTO 2014).

- From the workforce perspective, the fashion system's function is to provide a livelihood through selling raw materials, wages and salaries, to develop skills, etc.
- From the industry and business perspective, the function of the fashion system is to manufacture clothes in appropriate volumes to achieve profits, provide employment and contribute to a nation's economy (Allwood et al. 2006; Cao et al. 2008). Clothing manufacture has played a significant part in the development of a nation's economy; 58 % of global apparel exports are manufactured in developing country suppliers and it has been instrumental in the development of the East Asia's export growth and participation in the global economy (WTO 2014).

To understand the primary subsystem, the value chain concept is a useful framework (Gereffi 1999). A typical clothing and textiles value chain is organized into five parts, namely raw material supply which includes supply of natural and synthetic fibres, provision of components, production networks of garment factories, export channels by trade intermediaries and retail marketing networks. Progressing from one set of activities to the next requires some value-adding activity to make the process profitable (Cuc and Tripa 2008).

The global value chain for the textiles and clothing industry is often described as buyer-driven (Gereffi and Memedovic 2003). A buyer-driven value chain is one in which large retailers, marketers and manufacturers of branded items set up decentralized production networks in a number of exporting countries particularly the low wage ones. This networking has, itself, raised many social and environmental issues, which, unfortunately, are outside of the scope of this book to examine but nevertheless is an important consideration for a holistic study of the fashion industry.

Characteristics of the fashion industry may be noted as: time and cost, production networks and importance of design, branding and marketing and these are the elements that present barriers and opportunities for the fashion company that undertakes a remanufacturing ethos.

Cost
The apparel industry is driven by lowering time and cost. The proportion of total costs of production to manufacture needs to remain low in order to derive the maximum profit from the process. Clothing manufacture is still largely manual as there is still relatively no better fabric manipulation during the sewing process than the human hands. Described as a "beachhead sector" (where capabilities can be acquired easily), apparel manufacturing has become a "commodity" resulting in new divisions of labour and barriers for suppliers wishing to enter the value chain or to upgrade (Gereffi 1999; Bair 2005; Gereffi et al. 2001; Gereffi and Frederick 2010). Remuneration of manufacturing labour generally amounts to less than

one-tenth of the value of the final product, e.g. more than 95 % of personnel in the apparel value chain are employed in assembly line positions, mostly located in developing countries, and they receive less than 10 % of the product's value (WTO 2014). It is difficult for fashion companies engaged in remanufacture or upcycling processes to achieve parity in terms of costs against this mass manufactured fashion context due to the problems regarding access to volumes of fabric that would enable volume production (as already discussed earlier within this chapter and in Chap. 4). The problem with not achieving a comparable retail price for the garments would suggest that competing with high street market would be difficult without a good marketing strategy in place.

Time

The fashion market is characterized by short life cycles, high volatility, low predictability and high impulse purchasing (Christopher et al. 2004). Retailers need to spot trends quickly and translate them into the products for the shop floor in the shortest possible time and lowest cost so quickly imitate and produce only what sells in the market to sustain their competitive advantage (Christopher et al. 2004; Richardson 1996).

As many of the remanufacture and upcycle companies are design-driven and small in size, they have the capability to develop small collections that are ready for the market within a short time frame. The companies may manufacture themselves through a network of machinists employed either in the company or contracted as required. There are also a number of UK-based factories that can produce small runs of up to 50 units of trend-aware, trend-led or trend-setting garments within two or three weeks. It would appear, therefore, that the remanufacture fashion companies can compete with the large high street, mass market, retailers on the basis of timing if not volume or costs.

Global production networks

Apparel production for the mass market and high street takes place through globally distributed partners (McNamara 2008). Apparel production networks may include members in very distant (zone 3) or local/regional (zone 1, 1a) areas. Global retailers (which tend to have their origins and head offices within the West) with production in zone 3 areas (usually Asian) tend to have large suppliers. Retailers' high growth margins take advantage of low labour rates in zone 3 but the supply chains can take about 12 months to deliver a collection. Retailers who operate on the fast fashion model typically have a network of smaller suppliers at more local, regional, distance (zone 1 or 1a, i.e. Europe or Morocco). Here proximity to market enables the speed of response (2–3 weeks) but lower margins, as the labour rates are higher.

The "fast fashion" business model has been described as "lean retailing" with small batch production, short lead times and a reliance on near-shore production (Caro and Martinez-de-Albeniz 2015) which has helped the oft described "sunset industry" in developed economies (Abernathy et al. 2006; Doeringer and Crean 2006). Indeed, over the last few years, there has been a concerted effort by governments around the world to bring production back to their national shores in a process known as "reshoring" (Greenaway 2012). UK manufacturing is redeveloping (Gould 2014) and although it is not expected to reach the pre-1990s volume production, there are a number of manufacturing units that offer a range of services from basic cut-make and trim (CMT) to fully managed brand development services for companies that may be start-ups or more established; some are linked to larger production units overseas and can therefore can offer larger production runs at speeds to match the fashion market (DSA Manufacturing 2015; Sewport 2015). A number of free to use directories of UK clothing manufacturers such as www.letsmakeithere.org, www.stillmadcinbritain.co.uk, www.freeindex.co.uk have been developed to encourage production sourcing within the UK. For the remanufactured fashion companies, this is an opportunity that is playing to their particular needs—most of the companies are young, or start-up, a few are established and have developed a brand (e.g. Worn Again, Goodone, From Somewhere, People Tree). The production runs are still very small in comparison and there is a need to maintain price/cost balance; manufacturing at home may be more expensive but this has to be balanced by the costs of transport of goods, customs clearance and communication issues.

Importance of design, branding and marketing

Design, branding and marketing are crucial to the success of an apparel business, regardless of size or target market; calculated to add about 50 % to the whole value chain (Gereffi and Frederick 2010; UN 2010). Leading suppliers have developed both technical and design, marketing and branding capabilities to leverage competitive advantage and strengthen their supply chain relationships.

Caro et al. (2012) noted that the fast fashion model has three elements: (i) new product introduced throughout the season, (ii) demand typically decreases with time and (iii) new products are typically placed in more prominent positions to generate more interest than older ones. It has been noted that in the fast fashion business model design and marketing are strategic tools (Caro and Martinez-de-Albeniz 2015). Instead of the biannual approach to collections, the fast fashion retailers have taken a product-based view to assortment planning. To generate baseline sales and profits, they develop a classic/basic range that is produced in volume and which has a longer period of time on the shop floor to sell. At the same time, the retailers introduce new items with a high trend and design input throughout the season in small batches for a fast throughput; they are not replaced, or reduced in price, which generate further profit. Design is also a strategic asset for remanufactured/upcycled fashion companies which enables them to generate products in short time frames (as mentioned previously). While both have design as a strategic tool, they use this differently: the fast fashion retailer uses design to promote their modernity while the remanufactured/upcycled fashion companies use it to declare their design vision and individuality. This difference in using their design approach (apart from standardization issues) is also an area that warrants careful examination regarding selling through the retailers or developing capsule collections for the mass market, high street retailers.

(a)

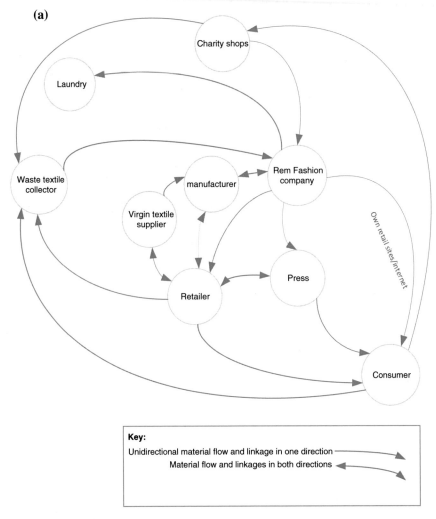

Fig. 5.5 **a** Current material flows within the supply chain and networks for the fashion industry. **b** Potential material flows within the supply chain and networks for the conceptual fashion manufacturing system (*Source* Authors)

5.4 A Conceptual System for Remanufactured Fashion

As already stated, it is vital for remanufacturers to build collaborative networks with established textile waste collection companies or develop product return systems, retailers and manufacturers. Figure 5.5a illustrates the material flows within the supply chain and networks for the fashion industry as currently operates. Figure 5.5b illustrates the material flows within the supply chain and networks that could and would need to be developed for the fashion industry to enable remanufacturing practice to be efficient.

(b)

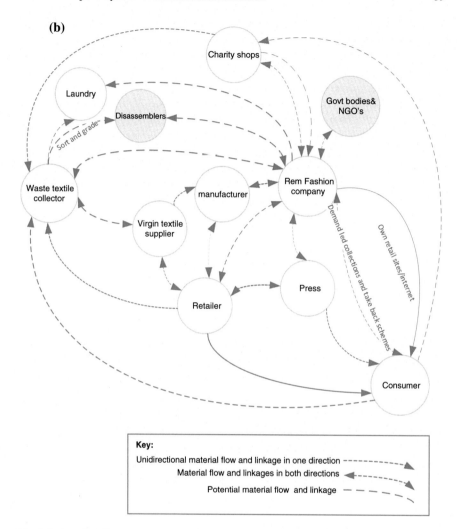

Fig. 5.5 (continued)

Waste textile collectors

Some design companies have developed supply networks with waste textile and second-hand clothing wholesalers to obtain a steady supply of input. Advantages of this for the design company include: (1) a steady supply of SHC, (2) increased likelihood of obtaining the required fabric types, colours and quantities depending on the production requirements, (3) minimizing waste generated due to in-house sorting processes and (4) minimizing the time spent on collection and sorting of SHC.

Waste collection within the fashion industry is still voluntary, and this is not so in other industries where some retailers have integrated waste collection activities into their operations. For example, for electronic and electrical equipment and cars, the producer takes the responsibility for product take-back and recycling, as detailed under the Extended Producer Responsibility (EPR) policy (Van Rossem et al. 2006). Some take-back systems already exist, such as those described within Chap. 3 of the EcoCircle and I:CO and reported elsewhere (Palm et al. 2014; Patagonia 2013).

The growth of reverse logistics channels in the remanufacturing business could be facilitated by retailer involvement in collecting waste. If the retailer takes the responsibility of taking the products back from the consumer and passing them to a waste collection company, it is highly likely the waste collectors would receive a significant volume of a particular style and/or a particular brand. The waste textile companies, as an add-on to their sorting process, could then launder or organize to send to launder and disassemble or perhaps new businesses could be developed that disassemble garments. This kind of a reverse flow could enable remanufacturing firms to obtain volumes of similar categories of clothing from waste collection companies. The SOEX Group, one of the world's largest textile waste collection and sorting companies, has developed relationships with manufacturers and retailers (e.g. Barneys New York) to restyle used clothing collected by SOEX Group (2012).

It would be possible for fashion retailers to develop relationships not only with waste collection companies, but also with remanufacturing firms in order to develop a closed-loop remanufacturing process and bring remanufactured fashion back to the store. Retailers could adopt this strategy as a part of their sustainable development plan which could facilitate material recirculation within the same loop and help develop local remanufacturing businesses.

Retailers
Current successful relationships have been developed with small boutiques, sustainable and ethical fashion shops. Where relationships have been developed with high street (mass market) retailers, it has been problematic due to retailer requirements for standardization, retail pricing strategies and volume of product (which many remanufacturing or upcycling designers are unable to maintain or supply). However, developing relationships with mass market retailers has advantages of: (1) developing brand awareness, (2) opportunity to develop efficiencies in collection of SHC, (3) consumer engagement with sustainable fashion and (4) maximizing the environmental impact of remanufactured fashion design process.

Manufacturing units
For remanufacturing companies, developing relationships with manufacturing facilities to outsource production enables: (1) redirect designers' time to developing new collections, (2) avoids the cost of setting up own manufacturing facility and (3) presents an opportunity to upscale production volumes.

Government and trade legislations bodies
As both the fashion and SHC industries are global, upscaling of remanufacturing processes often require outsourcing production in overseas locations where international trade and legal issues need to be considered as part of the process, in order to successfully manufacture and retail remanufactured ranges.

5.5   Marketing and Strategic Considerations for the Remanufacturing Company

Taking into account the design process of the remanufactured fashion company, an analysis of the strengths, weaknesses, threats and opportunities (SWOT) is presented in Table 5.2. The SWOT analysis identifies opportunities in several areas. Many companies already retail through own websites, or host retail company website or through a large fashion retailing company. Many also have their own brand label and some (such as People Tree or From Somewhere) are developing a brand presence in the more mainstream fashion market.

Table 5.2 SWOT analysis of the remanufactured fashion design process and company (*Source* Authors)

Strengths	Weaknesses

• Flexible design process
  Ability to supply more garments in small production numbers on request by retailers
  Ability to slightly amend designs on special request by retailer's customers
  Ability to incorporate last-minute decisions or changes (e.g. unavailability of fabrics)
• Production of one-off designs (special requests)
  Primary source of information about his consumer's needs and wants
• Personal vision of fashion is most important element of process
  Helps in creating a brand image
• Manufacturing situated close to studio
  Speed of manufacture and delivery
  Help develop employment within the sector
• Manufacturing part of company
  Greater level of control over the quality of production
• Focus strategy
  Concentrated effort to a small, defined group of potential customers
• Sustainability credentials
	• Very small number of permanent staff
	  Efficiency of production reduced
	  Time to conduct market and fabric research increased—adds cost
	  Time to treat sourced fabrics (launder or disassemble) increase—adds costs
	• Weak forward and backward linkages within the supply chain
	  Lack of control over the supply chain—e.g. lack of access to regular supply of fabrics or volumes of fabrics
	  Vulnerable to market fluctuations
	• Distribution through independent retailer
	  No control over product placement within the store
	  No regular contract for supply
	  Irregular payment
	  Not enough information about retail figures
	  Heavy reliance on retail buyers' knowledge of their consumers
	• No finances for promotion or advertising
	  Members of public remain unaware of the range
	• Lowered level of autonomy in comparison to regular designers
	  High authority over decisions about style/vision but counterbalanced by availability of colour and fabric
	• Difficult to convey or prove sustainability credentials
	• No quality control tests or measures

Table 5.2 (continued)

Opportunities	Threats
• Distribution/retailing through website Web-based retail or manufacturers with retail sites—orders are a primary source of consumer information • Small companies could develop cooperatives to consolidate on group's capabilities and, or, income and develop brands and PR • Brand collections—to develop consumer loyalty • Use textile designers to add to creative/unusual offering • Link with larger fast fashion model fashion retailers to deliver small collections with limited shop floor life • Links with government and NGO's to develop remanufacture label (quality assurance and sustainability credentials of the process) for funding and policy information exchange • Develop PR and educational materials for the consumer, e.g. swing tags, brochures, website campaigns	• Changing retail buying patterns • Very small market size Leads to small income and profit • Lack of finance difficult to overcome late payment • Lack of public awareness of brand may lead to loss of interest or enthusiasm for label

Fashion companies active in remanufacturing and/or upcycling (such as those described in Chap. 4) are small businesses and exhibit many of the characteristics of small businesses:

- Management is independent and managers also tend to be the owners
- Capital is supplied and ownership is often held by the individual, a group or family
- Operation is local (workers and owners are in one home community) but their customers and markets need not be local.

As the companies are small, they have to develop but strategically. Scott and Bruce (1987) explained their theory of small business development stages based on, among others, Greiner (1972) and also the product life-cycle theory. While the Greiner (1972) model was based on large company developments, the Scott and Bruce (1987) model is based specifically on observations of the small business and has five stages of development: inception, survival, growth, expansion and maturity. Each development stage is preceded by a form of crisis which will result in the company either folding or strengthened to develop to the next stage. It is possible to achieve stability without increase in size if that is the wish of the owner of the small business (e.g. if the business growth and objectives have been achieved to meet the goals and lifestyle objectives of the founder) termed as "comfort zone" (Ashworth 2010), or "contained growth" by Bruce and Scott (1987). How they choose to develop will have differing implications for the company. The remainder of this chapter will examine the literature review regarding the implications of these choices.

Growth into a large company

As noted earlier, many of the fashion companies are at the inception; business was founded on the founder's values and skills and all efforts are on product development with one manufacturing unit, a single market and limited channels of distribution. Crises faced at this stage are most often regarding finance (to progress from establishment to gaining profitability and cash flow), administration (recognizing the need for formalizing the systems and record keeping—perhaps a person has to be employed as not all entrepreneurs have these skills) and time (increased activity demands extra time to the business at the expense of family or personal time). Very often, the management role may be as supervisor and the organization structure may be unstructured—as the business grows, the need for more formalization requires a change in management style and role. In terms of product development, there is little research undertaken into the consumer or product; the product is often the reason for the business having been developed.

In Chap. 4, companies also displayed the survival stage where the company is potentially a stable business, and needs of the business centre on finance role to develop into working capital—i.e. the need to finance the increased inventories and receivables. Often this is gained through bank overdraughts and small loans. There is increased competition as more new companies embark on similar businesses if this is seen to be successful. The fashion industry has low barriers of entry, and therefore, this increase in competition requires the company to search for new channels of distribution. There is an increase in the number and types of crisis:

- Overtrading—this is the danger of accepting more orders for production than can be fulfilled, this is uncontrolled growth and requires careful handling to avoid.
- Increased complexity of expanded distribution channels—if successful in finding new channels of distribution, there are new customers, new values, new location and so new logistics; management styles have to be changed from supervision of staff to supervision of supervisors—i.e. employing more staff requiring greater delegation so losing some control over the process.
- Change in the basis of competition—with new competitors, there is a pressure on price points more than differentiation of products. If the emphasis is taken to go for the price-led competition, there is a requirement for volume production which requires financing placing more pressure on the management.
- Pressure for information—as competition grows, there are increased pressures for information about costs and cost controls require formalized control systems, perhaps through bringing in professional book keeping and auditors. This again leads to change in management style and potential changes in power base.

Scott and Bruce (1987) stated that companies may remain at the survival stage for some time, but they will be forced into taking the next stage (growth) either of their own volition or to meet the demands of the completion.

Growth has been defined as 15 % sales growth (Miller and Friesen 1984), but this has not necessarily been seen within the fashion small companies (Ashworth 2010). Companies exhibiting growth stage are profitable but not necessarily generating cash for the owner; any profits are returned to the company to consolidate on its position. The company is organized along functional roles and more time is spent on coordination of the functional managers. Professional accounting systems now in place, the company can undertake some research and development to expand their product range. If there are enough opportunities, the company can consider entering the expansion stage. However, the increased amount of competition will raise crisis of how to do this: compete on price or differentiation?

To compete on price requires an increase in market share and or new products, which will require financing (not just for growth but also for the operations required to produce the volume production). To compete on differentiation, expansion of market share is relinquished in favour of margin and product differentiation which requires more time and resources and so the company remains at this growth stage a longer time. The managerial style in this stage is more professional and somewhat decentralised, with increased delegation and relinquishing of some power, while the organizational structure is functional and decentralised.

A potential example of this is the case of Worn Again. Their story, presented on their website http://wornagain.info/, shows the development from an inception stage in 2005 producing shoes made from recycled materials and sold through Clark's shoe shops to developing into upcycling corporate uniforms into clothes and accessories. Having established their credentials as a business with a team of four people with roles from chief executive officer, to business innovations, researcher and financial controller as well as a board of four experts from the fashion industry as well as academia, the company now is researching into technology for separating and extracting fibres in mixed material garments. They are being supported in this venture by H&M (the fast fashion retailer who are now second in size to the Inditex group who control Zara brands) and Kering (the parent company of the brand PUMA) (Reuters 2015). It appears that Worn Again is entering the expansion/maturity stage. They are now capable of not only financing growth but also maintaining growth through access to funders (e.g. H&M and Kering). Moreover, the Worn Again website separates team members (with functional responsibilities as well as the day to day running of the company and strategic directions) from a board of directors (to help advise on strategic direction and funding sources) thus demonstrating that their management and organizational structure has become fully functional with formalized systems of control and regulation. Most importantly, research and development has moved from focus on product innovation and market research to one of production innovation, as their website states "product becomes system" (Worn Again 2015).

The crises for the companies in the growth, expansion and maturity phases are changes in management styles: the founder of the company becomes increasingly further away from the operations of the business and so in danger of being less involved or powerful in decision-making. The greater size of the company and increased competition also mean that the company now has to undertake greater research about the external environment and greater efforts into financing the marketing, maintaining and upgrading of the operations. Depending on the market, the founder may then sell the business or take it to the next stage which means coming away from being a small business to that of a corporation.

Controlled or Comfort zone growth

Some companies choose not to grow beyond a particular size, particularly if they do not wish to undertake production innovations as reported above. This "contained growth" requires a company to have achieved growth stage with all the characteristics listed as above. As there is an increasing number of design companies capable of remanufacture, there are opportunities for them to develop networks from strategic collaborations to partnerships to consolidate on the group's capabilities and/or income and develop brands and PR.

Caro and Martinez-de-Albeniz (2015) note that new models hinge on the design element of the product (although prices tend to be comparable to mass market). Market pull models use a form of open innovation and crowdsourcing approach such as at the threadless website which primarily prints on T-shirts. It encourages makers/artists to submit their own designs; the designs are voted for popularity; each week the most popular design is then printed on a T-shirt and sold. Every week new designs are chosen and the winning artist is paid for the design; there are also themed design challenges with larger cash prizes. There are other artefacts that can have the winning design printed on it (e.g. drinking flask, pillow cases, tote bags, laptop cases); the artist is paid for each time the print design is used. ModCloth, a clothing website, uses a similar model for manufacturing clothing, from designers who they retail as well as their own brand label. Market push platforms are evident in websites such as JustFab which curates hundreds of styles of shoes. The website asks the consumers to sign up to use the website, but in doing so they need to fill out what styles of shoes they prefer and so each day the website will email a set of shoe styles based on the preferences.

Other forms of strategic collaborations such as social enterprises can be found in organizations such as People Tree and Traid. People Tree was founded by Safia Minney, built on Fair Trade and organic cotton and wool. People Tree collaborates with a number of communities of craftspeople in the developing world such as the Indian subcontinent and Kenya's well as designers now household names (e.g. Zandra Rhodes, Orla Kiely). The craft communities often also support local schools, hospitals, etc. For each product, the People Tree website identifies who the item was made by and the payment structure, and how this benefitted the companies and communities involved. Profits made by the communities are often then put back into their areas of interest. TRAID was launched in 1999 as a charity that collected and resold clothing, with the aim of diverting clothing from the landfill. TRAID still resell second-hand clothing and the profits made are then put back into social projects around the world. In 2002, they launched the TRAIDremade fashion label which developed clothing made from waste textiles and collaborate with designers, artists and makers. Social enterprises use business methods to achieve social goals and have a variety of constitutional or legal forms: e.g. companies limited by guarantee or by shares; they may be mutual organizations (industrial and provident societies) or they may even be charities. The key difference from conventional private sector businesses is to maximize the reinvestment of any profit generated back into the achievement of social impact (Meadows and Pike 2010). Social enterprises include the following forms of organization:

community enterprises, cooperatives, the trading arms of charities, employee-owned businesses, development trusts, credit unions, housing associations and social firms (Vickers and Lyon 2014). According to Social Enterprise UK (2012), whatever form the social enterprise takes, there are some distinct characteristics in that they should have:

- Have a clear social and/or environmental mission set out in their governing documents
- Generate the majority of their income through trade
- Reinvest the majority of their profits
- Be autonomous of state
- Be majority controlled in the interests of the social mission
- Be accountable and transparent.

Social enterprises should generate at least 50 % of their income through trade activities and reinvest at least 50 % or more back into their social or environmental mission most often through supporting a charity. Funding for social enterprises in the UK has been through government schemes such as Social Venture Capital Fund, Adventure Capital Fund or Community Builders Fund. As funders are from private sector, therefore, the funding stream is a changing landscape that requires constant monitoring to ensure that funding is sought at the right time.

Whichever strategic choice is taken by the company (to grow with possible implications of the need to develop production innovations and possible changes in business model or develop contained growth through some form of strategic collaboration), there is a need for a company to achieve legitimacy. Company legitimacy for start-ups or new businesses has been described in the management literature as capturing the consumer market interests, convincing banks that they are a good idea to invest in and that they submit to the government rules, i.e. there is a social judgement of acceptance, appropriateness and desirability (Zimmerman and Zeitz 2002). This is an important resource to be within a company from the outset as it enables them to gain access to further resources crucial for any growth such as capital, technology, mangers, competent employees, customers and networks. This legitimacy can itself be enhanced by strategic choices of how it is gained. For the companies involved in remanufacturing, the objectives are for a sustainable fashion industry through making use of waste textiles—a process that is a departure from the current practice and the fashion system. It therefore requires considerable and pre-emptive "intervention" on the part of the company to develop bases of support in the society and the marketplace (Zimmerman and Zeitz 2002)—that is the norms and values of the fashion system and the fashion consumer have to be challenged to move away from the throwaway fast fashion to one of the more considered approaches to design the environment and the social structures involved in producing it. Some of the legitimacy can be gained through taking up environmental and social certification labels (discussed in Chap. 6) and also through developing links with the government and NGOs to develop a specific remanufactured clothing label that assures the buyer of its quality coupled with educational programmes not only marketing the label but also the processes involved and the benefits of remanufactured fashion.

5.6 Conclusion

This chapter has examined the current practices and supply chains in remanufactured fashion, identified the inefficiencies in its current practice and posited a conceptual system for remanufacturing companies. The concept of reuse and redesign is not new; this had been the normal method of clothes maintenance for centuries. Indeed, there was a thriving business of sorting and trade in used textiles since the 1600s; the world's largest trading centres were in the Netherlands and London, while the largest exporters were in the USA, the Netherlands and Belgium (Lemire 1991). In the UK, the wartime shortages and rationing of all resources during the Second World War made this a necessity (Atkinson 2006). The fashion system has been developing into its current complex form since the 1960s when youth no longer took the lead from the haute couture (Wilson 2009). To bring a concept born of necessity and frugality into the fashion sphere requires using knowledge about trends, the fashion cycle and the retailer.

Although the conceptual basis of remanufacture and fashion may be oppositional (the former of conserving, the latter of replenishment regardless of need), from a waste hierarchy perspective, remanufacturing makes sense. Fabric which has already been made is pulled out from the waste stream and used as a resource, avoiding supply chain activities that use chemicals and release hazardous waste to the environment, up to fabric production and saving energy and water. With regard to the environmental impact, recycling is less preferred than remanufacturing; however, remanufacturing is just one of a set of strategies that can be used for a more environmentally sustainable fashion industry.

To maximise the environmental benefits from remanufactured fashion, there needs to be increased consumption and availability. Increasing availability of remanufactured fashion may be achieved by either through increasing the number of design companies that remanufacture (which increases competition) or increasing the production capacity of current remanufacturing firms (which requires financing for production volumes and leveraging legitimacy).

In a discussion about the textile recycling system, Hawley noted that "...consumers must embrace the system...arbiters must continue to develop new value-markets and market the after-use possibilities" for a fully functional and effective system (Hawley 2006, p. 22). This is echoed by Niinimäki and Hassi (2011) who stated that a "new sustainable mindset is still waiting to emerge at large, as we continue to design and manufacture textile and clothing mainly in traditional ways" (Niinimäki and Hassi 2011, p. 1878), suggesting that more education and promotion about textile waste is needed.

References

Abernathy FH, Volpe A, Weil D (2006) The future of the apparel and textile industries: prospects and choices for public and private actors. Environ Planning A 38:2207–2232

Allwood J, Laursen SE, de Rodríguez CM, Bocken NMP (2006) Well Dressed? The present and future sustainability of clothing and textiles in the United Kingdom. University of Cambridge Institute for Manufacturing

Ashworth CJ (2010) The place of marketing in online fashion retail e-SMEs. EURAM conference, Rome, Italy, 19–22 May 2010

APSRG (2014a) Remanufacturing: towards a resource efficient economy. The All-Party Parliamentary Sustainable Resource Group, March, http://www.policyconnect.org.uk/apsrg/research/report-remanufacturing-towards-resource-efficient-economy-0

APSRG (2014b) Triple win: the economic, social and environmental case for remanufacturing. December 2014, http://www.policyconnect.org.uk/sites/site_pc/files/.../apsrgapmg-triplewin.pdf

Atkinson P (2006) Do it yourself: democracy and design. J Des Hist 19(1):1–10. doi:10.1093/jdh/epk001

Bair J (2005) Global capitalism and commodity Chains: looking back, going forward. Competition Change 9(2):153–180

Burns LD, Bryant NO (2002) The business of fashion: designing, manufacturing and marketing, 2nd edn. Fairchild Publications, Inc., New York

Cao N, Zhang Z, To KM, Ng KP (2008) How are supply chains coordinated? An empirical observation in Textile-apparel business. J Fashion Marketing and Management 12(3):384–397

Caro F, Martinez de Albeniz V (2015) Fast fashion: business model overview and research opportunities. In: Agrawal N, Smith SA (eds) Retail Supply chain management: quantitative models and empirical studies, 2nd edn. Springer, pp 237–264, ISBN 978-1-4899-7562-1

Caro F, Martinez-de-Albeniz V, Rusmevichientong P (2012) The assortment packing problem: multiperiod assortment planning for short-lived products. Working paper, IESE Business School

Charnley F, Lemon M, Evans S (2011) Exploring the process of whole system design. Des Stud 32(2011):156–179. doi:10.1016/j.destud.2010.08.002

Christopher M, Lowson R, Peck H (2004) Creating agile supply chain in the fashion industry. Int J Retail Distrib Manage 32(8):67–376

Cuc S, Tripa S (2008) Using the global value chain approaches in the clothing industries. Ann Oradea Univ, Fascicle Manage Technol Eng VII(XVII):2086–2090

Dissanayake DGK (2012) Sustainable and remanufactured fashion. Unpublished PhD thesis, University of Manchester

Doeringer P, Crean S (2006) Can fast fashion save the US apparel industry. Socio-Econ Rev 4:353377

DSA Manufacturing (2015) http://www.dsa-manufacturing.co.uk/

Fleischmann M, Boemhof-Ruwaard JM, Dekker R, van der Laan E, van Nunen J, Van Wassenhove LN (1997) Quantitative models for reverse logistics: a review. Eur J Oper Res 103(1):1–17

Forrester (1976) Business structure, economic cycles, and national policy. Futures 195–214 (Elsevier Ltd.)

Gereffi G (1999) International trade and industrial upgrading in the apparel commodity chain. J Int Econ 48:37–70

Gereffi G, Memedovic O (2003) Global value chain—Garment. UNIDO. http://www.unido.org/fileadmin/user_media/Publications/Pub_free/Global_apparel_value_chain.pdf

Gereffi G, Frederick S (2010) The global apparel value chain, trade and the crisis: challenges and opportunities for developing countries. Policy research, working paper

Gereffi G, Humphrey J, Kaplinsky R, Sturgeon TJ (2001) Introduction: globalisation, value chains and development. IDS Bull 32(3). https://www.ids.ac.uk/files/dmfile/gereffietal323.pdf

Greenaway (2012) The UK in a global world: how can the UK focus on steps in global value chains that really add value? Centre for Economic Policy Research (CEPR), ISBN 978-1-907142-55-0. http://www.nottingham.ac.uk/gep/documents/reports/dgebookcomplete.pdf

Gould H (2014) Returning fashion manufacturing to the UK—opportunities and challenges. The Guardian, 10 June. http://www.theguardian.com/sustainable-business/sustainable-fashion-blog/returning-fashion-manufacturing-uk-opportunities-challenges

Greiner LE (1972) Evolution and revolution as organizations grow. Harvard Business Review, July/August

Guide D, Van Wassenhove L (2001) Managing product returns for remanufacturing. Prod Oper Manage 10(2):142–155

Guide D, Teunter R, Van Wassenhove L (2003) Matching Demand and Supply to Maximise Profits form Remanufacturing. Working Paper Series. INSEAD

Hamilton JA (1987) Dress as a cultural sub-system: a unifying metatheory for clothing and textiles. Clothing Textiles Res J 6(*1):1–7

Hawley JM (2006) Textile recycling: a system perspective. In Wang Y (ed) Recycling in textiles. Woodhead Publications, London, pp 7–24

Kahhat R, Navia R (2013) A reverse supply chain can enhance waste management programmes. Waste Manage Res 31:1081–1084. doi:10.1177/0734242X13509722

Kaiser SB, Nagasawa RH, Hutton SS (1995) Construction of an SI theory of fashion: Part 1. Ambivalence and change. Clothing Text Res J 13(#3):172–183

Keiser SJ, Garner MB (2008) BEYOND DESIGN: the synergy of apparel product development. Fairchild Publications, Inc., New York

Krikke H, Pappis CP, Tsoulfas GT, Bloemhof-Ruwaard J (2001) Design principles for closed loop supply chains: optimizing economic, logistic and environmental performance. ERS-2001–62-LIS

Kumar N, Chatterjee A (2011) Reverse supply chain: completing the supply chain loop. Cognizant 20-20 insights, January. http://www.cognizant.com/InsightsWhitepapers/Reverse-Supply-Chain.pdf

Kunz G, Garner MB (2007) Going Global. The Textile and Apparel Industry, Fairchild

Lemire B (1991) Fashion's favourite: the cotton trade and the consumer in Britain, 1660-1800. Oxford University Press, Oxford

Majumder P, Groenevelt H (2001) Competition in remanufacturing. Prod Oper Manage 10(2):125–141

McKelvey K, Munslow J (2003) Fashion design: process, innovation & practice. Blackwell Science Ltd, Oxford

McNamara K (2008) The global textile and garment industry: the role of Information and communication technologies (ICTs) in exploiting the value chain

Meadows D (2008) In: Wright D (ed) Thinking in systems, a primer. Sustainable Institute, USA

Meadows M, Pike M (2010) Performance management for social enterprises: Systemic, Practice and Action Research, 23(2):127–141. http://dx.doi.org/doi:10.1007/s11213-009-9149-5

Michaud C, Llerena D (2006) An economic perspective on remanufactured products: industrial and consumption challenges for life cycle engineering. In: Proceedings of 13th CIRP international conference on life cycle engineering, 31 May–2 June 2006, Leuven, pp 543–548

Miller D, Friesen PH (1984) A longitudinal study of the corporate life cycle. Manage Sci 30:1161–1183

Mingers J, White L (2010) A review of the recent contribution of systems thinking to operational research and management science. Eur J Oper Res 207:1147–1161

Moore S, Wentz, M (eds) (2009) Eco-labeling for textiles and apparel. In: Sustainable textiles life cycle and environmental impact, Chap. 10. Woodhead Publishing, Textile Institute, Cambridge

Nasr N, Thurston M (2006) Remanufacturing: a key enabler to sustainable product systems. Rochester Institute of Technology

Niinimäki K, Hassi L (2011) Emerging design strategies in sustainable production and consumption of textiles and clothing. J Cleaner Prod 19:1876–1883. doi:10.1016/j.jclepro.2011.04.020

Palm D, Elander M, Watson D, Kiørboe N, Salmenperä H, Dahlbo H, Moliis H, Lyng H, Valente C, Gíslason S, Tekie H, Rydberg T (2014) Towards a Nordic textile strategy: collection, sorting, reuse and recycling of textiles. Nordic Council of Ministers. http://www4.ivl.se/downloa d/18.6cf6943a14637f76eab2776/1402668343852/C28+TN2014538+web.pdf

Patagonia (2013) Closing the loop—a report on Patagonia's common threads garment recycling program. The Cleanest Line. http://www.thecleanestline.com/2009/03/closing-the-loop-a-report-on-patagonias-common-threads-garment-recycling-program.html

Quinn FJ (1997) What's the buzz? Logistics Manage 36(2):43–47

Reuters (2015) Recycling—fashion world's antidote to environmental concerns. http://in.reuters.com/article/fashion-recycling-hm-idINL5N1100UM20150825. Accessed 25 Aug 2015

Richardson J (1996) Vertical integration and rapid response in fashion apparel. Organ Sci 7(4):400–412

Rosenau JA, Wilson DL (2006) Apparel merchandising: the line starts here, 2nd edn. Fairchild Publications, Inc., New York

Savaskan RC, Bhattacarya S, Van Wassenhove LN (2004) Closed-loop supply chain models with product remanufacturing. Manage Sci 50(2):239–252

Scott M, Bruce R (1987) Five stages of growth in small businesses. Long Range Planning 20(3):45–52

Seadon JK (2010) Sustainable waste management systems. J Cleaner Prod 18:1639–1651. doi:10.1016/j.jclepro.2010.07.009

Sewport (2015) http://www.sewport.co.uk/

Smithers R (2012) TK Maxx and Cancer Research team up for children's cancer research. The Guardian, 30 Mar. http://www.theguardian.com/money/blog/2012/mar/30/tkmaxx-cancer-research-donations

Sinha P (2000) The role of designing through making across market levels in the UK fashion industry. Des J 3(3):26–44

Sinha P, Hussey C (2009) Product labelling for improved end-of-life management: an investigation to determine the feasibility of garment labelling to enable better end-of-life management of corporate clothing. Centre for Remanufacturing and Reuse, Oakdene Hollins, DEFRA Clothing Roadmap Study. http://www.uniformreuse.co.uk/futureindex.html

Sinha P, Tipi N, Beverley K, Schultz R, Day C, Domvoglou D (2014) Waste textiles as raw materials in sub-saharan Africa: supply chain challenges. LRN annual conference 2014, University of Huddersfield, 3-5 Sept, ISBN 978-1-904564-48-5

SOEX Group (2012) Re-collection. Available at http://www.soexgroup.com/#/Partners_&_Affiliates/Re-Collection/

Stock JR (2001) Reverse logistics in the supply chain. Business briefing: global purchasing and supply chain management, Illinois, pp 44–48

Studd R (2002) Textile design process. Des J 5(1):36–49

Sundin E (2004) Product and process design for successful remanufacturing. Published doctoral dissertation, Linköping's University, Sweden

Van Rossem C, Tojo N, Lindhqvist T (2006) Extended producer responsibility: AN examination of its impact on innovation and greening products. Greenpeace international, friends of the earth and the European environmental Bureau (EEB)

Veiga MM (2013) Analysis of efficiency of waste reverse logistics for recycling. Waste Manage Res 31:26

Vickers I, Lyon F (2014) Beyond green niches? Growth strategies of environmentally-motivated social enterprises. International Small Business Journal 32:449–471

von Bertalanffy L (1950) An outline of general system theory. Br J Philos Sci 1:134–164

Wilson E (1985, 2009) Adorned in dreams: fashion and modernity (revised edition). I B Tauris & Co Ltd

Worn Again (2015) www.wornagain.co.uk

WTO (2014) International trade statistics. https://www.wto.org/english/res_e/statis_e/its2014_e/its2014_e.pdf

United Nations (2010) Creative economy report. www.Unctad.org/creative-economy

Zimmerman MA, Zeitz GJ (2002) Beyond survival: achieving new venture growth by building legitimacy. Acad Manage Rev 27(3):414–431. http://www.jstor.org/stable/4134387

Chapter 6
Issues Raised for Sustainability Through Remanufactured Fashion

6.1 Retailing of Remanufactured Fashion

This final chapter returns to our original issue of defining remanufactured fashion—that there is currently no certification specific to the process of remanufacture within the fashion industry. Certification takes on significance when viewed in the contexts of recent research that noted that although 81 % of a sample of 4000 men and women had bought at least one item of second-hand clothing, concerns over quality of the garment, a lack of guarantee and lack of durability prevented them from buying second hand more regularly (WRAP 2015). Chapter 5 discussed the need for legitimacy and one of the ways in which to do so is to evidence some compliance with industry and government standards. Alongside the quality requirements of the buyers, apparel manufacturers need to meet government and market standards. The EU buyer requirements (CBI 2014) for apparel have been categorized into three areas:

- Must—legal obligation to meet standards such as product safety, non-use of hazardous chemicals, CITES (declaration of products made from wild animals or plants) and labelling to identify fibre composition.
- Common—standards and labelling those competitors in the same market might have—e.g. company processes and practices that have achieved a standard of environmental awareness as judged by their ISO1400 approval or socially responsible through the award of SA 8000.
- Niche—voluntary standards that may help the product compete against others within its class—e.g. eco-labels that require validation through certification.

© Springer Science+Business Media Singapore 2016
P. Sinha et al., *Remanufactured Fashion*, Environmental Footprints
and Eco-design of Products and Processes, DOI 10.1007/978-981-10-0297-7_6

6.2 Eco-Labels

Because of the environmental objectives of remanufactured fashion (e.g. diverting textiles from landfill, resource maximization), a certificate of remanufacture might be considered as an eco-label. To examine if this is appropriate, we consider how eco-labels are used within the textiles and fashion industry.

Eco-labels are used by companies for conveying environmental information about a product to the consumer and communicate that the environmental impacts are reduced over the entire life cycle of a product without specifying the production practices (Piotrowski and Kratz 2005). According to the European Union, the role and objective of the eco-label in the Integrated Product Policy (IPP) is to provide better information to consumers about a product or series of products and its environmental and/or social impact (European Commission 2009). The IPP aims to minimize the environmental degradation caused by any of the phases of a product life cycle (tangible or intangible, such as service), e.g. manufacture, development, use or disposal (European Commission 2008). All phases of a product life cycle are examined with the objective of improving their environmental performance. This approach requires all participants in this process to be engaged, e.g. designers, industry, marketing managers, retailers and consumers.

There are two motives driving eco-labelling in the textile industry: (i) the environmental safety and (ii) improving working conditions (Golden 2010). Eco-labels are voluntary, and the International Standards Organization (ISO) classifies them into three types I, II and III (EPA 1998).

- Type I labels are third-party certified and use a logo to communicate that the criteria have been met.
- Type II labels are self-declared by manufacturer, distributor, etc.
- Type III labels provide the full quantitative LCA data as a report.

Eco-labels are normally issued either by government-supported or private enterprises once it has been proved that the product has met the criteria (Hyvarinen 1999). Examples of government-backed schemes are as follows: Blue Angel (Germany), Eco Mark (Japan), Environmental Choice (Canada), White Swan (Nordic Countries), EU, Eco-Mark (India) and Green Label (Singapore). Examples of schemes developed by companies, laboratories or consortiums outside of government ownership include the following: Eco-tex and Oeko-Tex (textiles and clothing—Germany) and Green Seal (USA). Criteria for granting eco-labels are mostly based on the "cradle-to-grave" approach, i.e. the life-cycle analysis of the product and assessment of its impact on the environment from processing of raw materials, production, distribution, consumption and maintenance, (i.e. washing, ironing, dry-cleaning) and finally disposal of the product. A "Cradle to Cradle" certification programme assesses the sustainability of product ingredients for human and environmental health, as well as their recyclability or compostability making it easier at the design stage to create ecologically intelligent products through choosing materials that meet key sustainability criteria for material health and material reutilization (Braungart and McDonough 2008). The Global Recycled Standard (GRS) and R Cert as described in chapter one are two examples of this approach. The certification period varies between 1 and 5 years; at the end of this, the certification process has to be undertaken again to revalidate the use of the eco-label. The eco-label provider is assigned the authority to undertake "spot" checks during any time of the valid period of the eco-label. Certification provides a comprehensive system for ensuring that certain standards of production and processing are met, such as developing rules or standards (standard setting), verifying and evaluating performance against those standards (inspection) and recognizing procedures which successfully meet the standards (certification). Two kinds of certificates are required in the textiles and fashion supply chain (Shah 2010):

- Scope: issued to a supplier stating its name, the products and production facility inspected and certified in accordance with standards.
- Transaction: issued to the seller for the sale of products if the applicant has the scope certificate.

A retail buyer may require a scope certificate from a chosen garment manufacturer to prove that their manufacturing facility is certified and products are being produced in accordance with standards. If the manufacturer is not certified, they contact the certification body for the certification. If the garment manufacturer does not manufacture textiles, then the retail buyer would need to source the required certified raw material to process into fabric and factories certified and able to provide a scope certificate to garment manufacturer (if there is no certificate, then they need to contact the certification body to certify their facility). If the ordered garment has embroidery and printing, then either the garment manufacturer should acquire certification for those facilities under their scope, or suggest to the embroiderer or printer to acquire certification themselves through using certified threads, dyes or chemicals. Once all the relevant scope certificates have been obtained, the buyer (retailer) may then apply for a transaction certificate (Shah 2010).

The entire supply chain requires consistency in certification bodies used, i.e. if the garment manufacturer is required to produce to GOTS standards, the processors must also have GOTS certification—a different certification will not suffice as they each have their own analysis methods and processes for certification. Obtaining a certificate can take between 30 and 45 days, depending on the standards, the factory condition, the number of factories under one application, the understanding of standards by applicant and changes required within the factory with respect to compliance with standards. This process continues "backwards" along the supply chain to raw material supplier (e.g. organic cotton farmer to spinner). If during an inspection any non-compliance occurs, the process would be lengthened as each certification takes about 30–45 days (Shah 2010).

An important factor in the success of any eco-labels is its ability to cover its certification costs and therefore stay in business. Cost of certification can be high in relation to the value of the product and thus can become prohibitive. This is especially true for textiles because of the number of process involved from production to consumer. The expenses incurred in obtaining eco-certification can be especially prohibitive to companies engaged in remanufacture or recycle as they generally tend to be small to micro enterprises. The EU defines micro-enterprises as employing fewer than 10 people with annual turnover and/or annual balance sheet total not exceeding EUR 2 million (EUR-LEX 2007). The EU Ecolabel website currently states that, for micro-enterprises, the costs can be in the following region: an initial fee of €200–350 plus an annual renewal fee of up to €18,750 (EC Europa 2015). The exact prices vary depending on regional offices that are organizing the certification. These costs become greater down the chain as the costs of the labelled textiles are incorporated into the product development process.

Although certification is a complex, time-consuming and costly process, it creates transparency within the supply chain. For each step in the supply of a product, the transaction certificate provides assurance to buyer about the compliance of products manufactured in accordance with the standards of the certification or eco-label in question.

6.3 Remanufactured Certificates

Although it can be argued that remanufacture has environmental benefits, current certificates for proof of remanufacture are not a statement about the environmental qualities of the process, but rather they attest to the (re)manufacturing qualities. For instance, the certifications reviewed (Ford 2015; Breville 2015; ISO 2015) provide the following:

- Warranty of the product manufacture claims over a period of time set by the manufacturer
- Only the manufacturer or an approved agent of the manufacture can carry out the remanufacture
- The remanufactured product, if it fails, will be replaced in whole or repaired by the manufacturer or approved manufacturer's agent
- There is often an external testing house that conducts the tests and provides a certificate that states that the appropriate manufacturing quality specifications have been achieved. Indeed, there is ISO 13534:2000 specifically for remanufacturing of drilling and production equipment in the petroleum and natural gas industries.

Given the heavy reliance on the manufacturer, there needs to be an agreement on the (re)manufacturing quality standards to be used. Although there are accepted quality measures within the fashion industry, the standards are not uniform across the industry—market level dictates the retail price and therefore the manufacturing standards to strive for. The potential areas for consideration of the fashion remanufacturing process may be simply illustrated as in Fig. 6.1. The main question with regard to developing fashion remanufacture for the mass market level is that of costs versus benefits. Would a remanufacture certificate benefit a large mass market retailer in developing products with added value in partnership with a remanufacturing fashion company?

For instance, the increased uptake of eco-labels by textile manufacturers in the Indian subcontinent had been attributed as being of strategic importance to their companies (Austgulen et al. 2013; Shah 2010), especially with respect

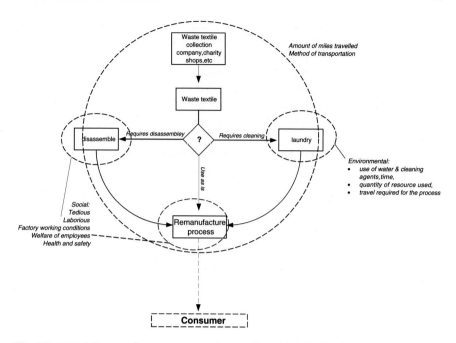

Fig. 6.1 Potential remanufacture process and areas of consideration for the certification process (*Source* Authors)

to promoting their corporate social responsibility, adding competitive advantage and strengthening their image and reputation. The study by Austgulen et al. (2013) also identified that where the Indian producers had complied with global eco-labels such as GOTS, their export volumes to Europe increased. This has significant implications regarding how the trade in eco-labelled products develops: many Indian textile exporters and global apparel retailers stated a preference for global eco-labels, while firms with internal markets found that consumers knew and trusted the local eco-label but did not know the global ones.

Marketing and the consumer confusion?

As the apparel industry is customer driven, eco-labels have been regarded as an important tool in communicating and developing sustainable consumption with people, process and profit in mind. The USA EPA (1994) defined the following five factors for measuring effectiveness of an eco-label, the first four of which serve to support the last: (i) consumer awareness of labels, (ii) consumer acceptance of labels (credibility and understanding), (iii) changes in consumer behaviour, (iv) changes in manufacturer behaviour and (v) net environmental gains. However, in reality, it evades examination because of the complexities in isolating their impacts from other measures (Austgulen et al. 2013; Golden 2010; OECD 2005).

Although it is difficult to assess the proportion of the global industry that has acquired an eco-label, Golden (2010) found that the most successful eco-labels were as follows: Oeko-Tex (health and safety), GOTS (organic textiles) and

FairTrade (social and labour conditions). These labels cover the entire supply chain thus: (i) negating the need for smaller labels, (ii) reducing consumer confusion and (iii) developing supply chain networks that are working within the same label criteria. These are also global labels, involve expert communities from across the world, setting standards that exceed those in home nations, and overcome politicized debates within countries to achieve trust.

Eco-labels are not simple to understand. For example, GOTS and OCS (Organic Content Standard) that replaced OE 100 from 2013 standards labelling guide are both standards applicable for products made from organic cotton. However, GOTS is based on social, technical and environmental areas, while the OCS is for tracking and documenting the purchase, handling and use of the amount of organically farmed cotton fibres (or organic-in-conversion cotton fibre) in yarns, fabrics and finished goods. However, OCS logoed garment may be made from 100 % organically grown cotton but then finished with harmful dyes and chemicals, printed with non-eco-friendly printing technique like solvent- based printing.

For recycled products, GRS claims the amount of recycled materials in the final product and also covers the issues pertaining to the environmental and social criteria, raising the potential to develop a similar type of a standard for remanufactured products. For example, a remanufacturing standard could verify the percentage of reused material that replaced virgin material that would otherwise have been used to make the garment. If the product is entirely made from waste material, the product could be claimed as 100 % remanufactured. However, if the percentage amount of waste material varies from product to product, to develop a label that claims the content of waste materials being reused would be a complex issue. Moreover, if the product is claimed as remanufactured through a certain standard, the process needs to be standardized beforehand and the audit procedures need to be defined and developed. This would be a long-term strategy for remanufactured fashion as the business grows, yet an eco-label that conveys the key message to the consumer is a current necessity.

The lack of common definition or general understanding for what constitutes environmentally friendly clothing and eco-labelling is a problem in marketing communication. The significance of this is lost on the consumer who is largely unaware of the differences. Moreover, of the 457 eco-labels identified worldwide, 108 cover textiles (Ecolabel Index 2014), a 263 % increase over the last six years when there were 47 identified (ecolabelling.org 2008). The proliferation of voluntary certification and labelling schemes for environmentally and socially responsible production is often seen as driven by companies and consumers. Consumers are heavily involved in environmental pollution because of their buying behaviour and consumption of textiles; influencing factors on consumers' willingness to buy environmentally friendly products have been identified and categorized as follows: demographics, knowledge, values, attitudes and behaviour (Laroche et al. 2001). However, as noted in chapter one, price has been found to be one of the most decisive factors in determining when consumers actually purchase apparel products. Consumers' willingness to pay and purchase cloths made from sustainable raw material such as organic cotton is a complex issue (Gam et al. 2010).

Much of the remanufactured fashion is developed by fashion design studios with fashion brands' labels which may indicate that they have a history in or are specialists in upcycling and/or remanufacturing. It is reasonable to question the necessity of the remanufacture certification; however, as described in chapter five, companies may change strategic direction, or their product ranges and the use of remanufacture certificate may help to add value and differentiation to the ranges offered. Further questions that are also raised and that need to be addressed may be noted as follows:

- Should quality tests for remanufactured fashion be standardized?
- What quality tests should be a part of the certificate?
- Who would be involved in developing the certification?
- Should the certificate be self-certification or third party?
- Should the certificates incorporate the source of the textiles used?
- Should social as well as environmental test or certification form part of the remanufacture certificate (as some of the process will undoubtedly be vulnerable to malpractices—see Fig. 6.1)?
- Should the certificates include distance travelled and methods of travel?
- Should certificates incorporate any use of laundry (water and chemicals)?

Determining the carbon footprint of the fashion remanufacturing industry would be an interesting aspect. The secondary textile and apparel industry claims to be a sustainable industry, however, there is a significant amount of transportation (and associated carbon footprint) involved in the collection of SHC, transporting them to sorting centers and then redistributing them to local or global resell markets or recyclers. The Framework Directive on Waste (75/442/EEC) established fundamental principles of managing waste in Europe, which must be reflected in national, regional and local strategies. "Regional self-sufficiency" means that most waste should be treated or disposed of within the region it is produced (Waste Strategy 2009). Each region is responsible for providing sufficient facilities and services to manage the waste produced within the region. The "proximity principle" states that waste should be managed as close as possible to where it is produced (Waste Strategy 2009). The main purpose is to limit the environmental impact on transporting waste and to create a more responsible approach to waste generation. Current upcycling and remanufacturing businesses support the "proximity principle", as they treat the waste garments and textiles within the region that it is generated. However, to develop into an environmentally conscious industry, the collection and redistribution strategies may need to be revised to reduce the transportation issues. The process of waste collection in the fashion industry differs from those manufacturing industries where government legislations have driven waste collection activities to be integrated into their operations. For example, for electronic and electrical equipment and car manufacturers, the producer takes the responsibility for product take-back and recycling, as detailed under the Extended Producer Responsibility (EPR) policy (Van Rossem et al. 2006). The textile industry currently has no such initiative and so waste textile collection is usually left out of main activities. Fashion retailers have a significant role to play to reduce extensive transportation involved in door-to-door collections and also to initiate the reverse flow of goods. SHC collection centres could be located near retailing stores or large fashion malls to facilitate an efficient take-back systems. As per the proximity principle, retailer owned waste collection and remanufacturing facilities would assist the regional management of waste that is being produced in the same region.

Remanufacturing is a sustainable business model that could be operated in any country. As massive amount of wastes are being shipped to developing countries at the moment for direct reuse, it has raised many implications in the destination markets including an oversupply of SHC, increasing competition, price reduction and undermining the local textile and clothing industries (Dupin 2003; Claudio 2007). The SHC markets are becoming saturated, and the social, economical and environmental impacts of the oversupply of SHC are not clearly understood (Dissanayake and Sinha 2012). In order to overcome some of those issues, a remanufacturing business model in the destination countries could be proposed. The excessive SHC would be remade to local, traditional dresses with the support of local designers and craft people. Reprinting the fabrics to suit the local community, use embroidery or other handcrafted techniques are some of the ways of providing a fresh look to the old materials. This would be an opportunity to build up a local apparel industry by creating new employment and markets for local apparel products and feeding some money back to the economy. While there is a potential that the products with local heritage could be exported back to the Western consumers, the environmental impact associated with the global transportation would be an issue.

6.4 Conclusions and Perspectives

The apparel industry has no legal pressures on sustainability from the governments, beyond the measures mentioned above. Governments are encouraging and helping industry develop voluntary measures to regulate themselves or encourage sustainable consumption within the marketplace. Initiatives such as the US-based Sustainable Apparel Coalition have developed an educational tool (the Higg Index, http://www.apparelcoalition.org/higgoverview/) to help industry members measure and rate their processes. This tool aims to help manufacturers, brands, NGO's and educational institutes become familiar with the language of this area and understand the implications of their processes better, and paid-up members can share best practice. The UK fashion industry, with WRAP, have developed a voluntary agreement called the SCAP 2020 Commitment which is a commitment to reduce the environmental footprint of a supply chain. All areas of the fashion industry (e.g., retailers, textile waste collectors, education) are encouraged to sign up to this agreement, the first commitment of which is to achieve a 15 % reduction in production of carbon, waste and use of water by 2020. To help signatories measure their performance a tool has been developed (The SCAP Footprint Calculator), which measures the use of lower impact fibres, extending the active life of clothes, and amount of reuse and recycling practices (WRAP 2014).

There is a role for sustainable venture capital (SVC) funders in the development of sustainable venture start-ups and young companies. Bocken's research (2015) found that key success factors when SVC was involved in a sustainable enterprise were business model innovations, collaborations and strong business cases and conversely key failure factors in sustainable enterprises were lack of suitable investors, a strong incumbent industry and short term mindsets. The advice for the sustainable entrepreneur was to focus on their triple bottom line in any business innovation, find new opportunities in technology and funding platforms and develop multiple business cases for success beyond the "green consumer"; the advice for the SVC was to coinvest and develop patience to balance financial returns with social and environmental ones (Bocken 2015).

References

Austgulen MH, Stø E, Jatkar A (2013) The dualism of eco-labels in the global textile market. In: An integrated Indian and European perspective" paper for OrganizaciónInternacionalAgropecuaria S.A. (OIA), 17 June 2013 http://www.oia.com.ar/documentos/ecolabels.pdf

Bocken NMP (2015) Sustainable venture capital e catalyst for sustainable start-up success? J Cleaner Prod 1–12 (in press). http://dx.doi.org/10.1016/j.jclepro.2015.05.079

Braungart M, McDonough W (2008) Cradle to cradle: remaking the way we make things. Vintage Books, London

Breville (2015) http://remanufactured.brevilleusa.com/blogs/news/9332693-warranty

CBI (2014) "EU buyer requirements for apparel" CBI, Ministry of Foreign Affairs of the Netherlands, Market Information Database, http://www.cbi.eu/marketintel_platform/apparel/135943/buyerrequirements

Claudio L (2007) Waste couture: environmental impact of the clothing industry. Environ Health Perspect 115 (9):448–454

Dissanayake DGK, Sinha P (2012) Sustainable waste management strategies in the fashion industry sector. Int J Environ Cult Econ Soc Sustain 8(1):77–90. The sustainability collection. ISSN 18322077

Dupin C (2003) The shirt off our backs. J Commer http://www.accessmylibrary.com/coms2/summary_0286-7971787_ITM

Ecolabel Index (2014) www.ecolabelndex.com

EC Europa (2014) How to apply for EU ecolabel, http://ec.europa.eu/environment/ecolabel/how-to-apply-for-eu-ecolabel.html

European Commission (2008) What is integrated product policy? http://ec.europa.eu/environment/ipp/integratedpp.htm

European Commission (2009) Report on the state of implementation of integrated product policy, the commission to the council, the European parliament, the European economic and social committee and the committee of the regions, SEC(2009)1707, http://eur-lex.europa.eu/legal-content/EN/TXT/?uri=CELEX:52009DC0693EUR-LEX (2007)

EPA (Environmental Protection Agency) (1994) Determinants of effectiveness for environmental certification and labelling programs. US EPA, Washington

EPA (Environmental Protection Agency) (1998) Environmental labelling, issues, policies, and practices worldwide, pollution prevention division office of pollution, prevention and toxics. USA

Ford (2015) Ford certified pre-owned limited warranty, http://www.primefordauburn.com/Media/Default/Page/ford_cpo_warranty.pdf

Gam HJ, Cao H, Farr C, Kang M (2010) Quest for the eco-apparel market: a study of mothers' willingness to purchase organic cotton clothing for their children. Int J Consum Stud 1–9

Golden JS (ed) (2010) An overview of ecolabels and sustainability certifications in the global marketplace. Duke University, North Carolina, USA, http://center.sustainability.duke.edu/sites/default/files/documents/ecolabelsreport.pdf

Hyvarinen A (1999) Eco-labelling and environmentally friendly products and production methods affecting the international trade in textiles and clothing. Senior Market Development Officer, International Trade Centre, Geneva

ISO (2015) ISO 13534:2000(en), petroleum and natural gas industries—drilling and production equipment—Inspection, maintenance, repair and remanufacture of hoisting equipment, https://www.iso.org/obp/ui/#iso:std:iso:13534:ed-1:v1:en

Laroche M, Bergeron J, Barbaro-Forleo G (2001) Targeting consumers who are willing to pay more for environmentally friendly products. J Consum Mark 18(6):503–52

OECD (2005) effects of eco-labelling schemes: compilation of recent studies, JT00179584, http://www.oecd.org/officialdocuments/publicdisplaydocumentpdf/?doclanguage=en&cote=com/env/td(2004)34/final

Piotrowski R, Kratz S (2005) Eco-labelling in globalised economy. In: Pfaller A, Lerch M (eds) Challenges of globalization, Ch. 13. Transaction Publishers, New Jersey

Shah R (2010) The role of different actors within the textiles—fashion supply chain to understand the issues regarding growth eco-labelled sustainable textiles products (STP). Unpublished MSc by research thesis, University of Manchester

Waste Strategy (2009) Waste management strategy for bexley, 2009–2014. Available at: www.bexley.gov.ukthe Higg Index, http://www.apparelcoalition.org/higgoverview/

Van Rossem C, Tojo N, Lindhqvist T (2006) Extended producer responsibility; an examination of its impact on innovation and greening products. Greenpeace International, Friends of the Earth and the European Environmental Bureau (EEB)

WRAP (2014) Clothing sector agrees challenging targets to cut environmental impact by 15 % http://www.wrap.org.uk/SCAP2020targetstraderelease

WRAP (2015) Study into consumer second-hand shopping behaviour to identify the re-use displacement effect http://www.wrap.org.uk/sites/files/wrap/Re-use%20displacement%20report%20executive%20summary%20v2.pdf